COLLECTIVE EXCELLENCE

Building Effective Teams

MEL HENSEY

SECOND EDITION

American Society of Civil Engineers
1801 Alexander Bell Drive
Reston, Virginia 20191-4400

Abstract: *Collective Excellence: Building Effective Teams, 2E* examines the team approach in the workplace as effective team development is an essential element in the successful operation of any organization. It presents the concepts and practices of team development and management for optimum productivity, moving systematically from defining teams and the various stages of team development to providing measures of team strengths and methods for capitalizing on each member's personality and temperament. Separate chapters address project management and the concept of leadership.

Library of Congress Cataloging-in-Publication Data

Hensey, Mel.
    Collective excellence : building effective teams / Mel Hensey.--2nd ed.
        p. cm.
    Includes bibliographical references and index.
    ISBN 0-7844-0546-8
        1. Teams in the workplace. I. Title.

HD66 .H46 2001
658.4'02--dc21   2001037319

Any statements expressed in these materials are those of the individual authors and do not necessarily represent the views of ASCE, which takes no responsibility for any statement made herein. No reference made in this publication to any specific method, product, process, or service constitutes or implies an endorsement, recommendation, or warranty thereof by ASCE. The materials are for general information only and do not represent a standard of ASCE, nor are they intended as a reference in purchase specifications, contracts, regulations, statutes, or any other legal document. ASCE makes no representation or warranty of any kind, whether express or implied, concerning the accuracy, completeness, suitability, or utility of any information, apparatus, product, or process discussed in this publication, and assumes no liability therefore. This information should not be used without first securing competent advice with respect to its suitability for any general or specific application. Anyone utilizing this information assumes all liability arising from such use, including but not limited to infringement of any patent or patents.

ASCE and American Society of Civil Engineers—Registered in U.S. Patent and Trademark Office.

Photocopies: Authorization to photocopy material for internal or personal use under circumstances not falling within the fair use provisions of the Copyright Act is granted by ASCE to libraries and other users registered with the Copyright Clearance Center (CCC) Transactional Reporting Service, provided that the base fee of $8.00 per article plus $.50 per page is paid directly to CCC, 222 Rosewood Drive, Danvers, MA 01923. The identification for ASCE Books is 0-7844-0546-8/01/ $8.00 + $.50 per page. Requests for special permission or bulk copying should be addressed to Permissions & Copyright Dept., ASCE.

Copyright © 2001 by the American Society of Civil Engineers.
All Rights Reserved.
Library of Congress Catalog Card No: 2001037319
ISBN 0-7844-0546-8
Manufactured in the United States of America.

ABOUT THE AUTHOR

Mel Hensey is Carol Hensey's partner in their management consulting firm in Maineville, Ohio, near Cincinnati. Their work focuses on helping public and private organizations with projects such as:

- strategic planning,
- organization redesign,
- leadership development,
- organizational problem solving, and
- executive team building.

Mel enjoyed working in several technical organizations as an engineer and manager before he and Carol formed their consulting group in 1974. His work in Earth Science Labs (consulting), Cincinnati Bell (construction), and Procter & Gamble (engineering) provided the practical experience needed to serve his clients' needs.

Clients with whom the Henseys have recently worked include LJB; BWSC; Investment Scorecard; Danis Building Construction; the Children's Hospital of Cincinnati; Terracon Consultants; Ferro Corporation; the Universities of Cincinnati, Missouri, and Toledo; as well as Drexel, Miami, Purdue and Texas A&M Universities; ASCE; ASFE; the U.S. Army Corps of Engineers; and the cities of Cincinnati, Cleveland, and Louisville.

Mel has served on the faculty of several management institutes, including the Construction Executive Program of Texas A&M University and the Executive Effectiveness Course of the American Management Association.

As founding editor of ASCE's *Journal of Management in Engineering*, Mel guided the publication through its first six years. Shortly after receiving the Torrens Award from ASCE (1989), he turned the journal over to a team of colleagues who have steadily improved it.

Carol and Mel enjoy traveling, canoeing, antiquing, volunteer work, and their four children and six grandchildren. Their firm has a family team flavor:

- Mel is principal consultant.
- Carol is the office and financial manager.
- Ann provides secretarial services (Professional Office Services, Fairfield, Ohio).
- Chris, Terry, and young Mel have all helped over the years!

ACKNOWLEDGMENTS

My own learning and appreciation for the natural advantages of teams and teamwork have come from hundreds of sources. These include several friends and colleagues who nearly "broke their picks" trying to make me more of a team player.

The most tangible and exciting evidence of the magic of teamwork came from my participation in our children's camping, canoeing, and scouting adventures—both those that went well and those that developed the inevitable difficulties. In our early version of "Outward Bound," we six learned a lot about teams!

My colleagues in the Engineering Division of Procter & Gamble made many contributions prior to 1974. Fine models of team leadership were provided by Jim Golan, Ray Carlin, Bill Richards, and Mike Pedicini. Supportive team membership was also an important contribution from many P&G colleagues, including Betty Kloecker, Ed Marcotte, Bob Harrison, and Ray Rose.

The organizations it has been my privilege to serve as consultant—for planning, team building, reorganization, and such—have also taught me a great deal. Far too numerous to list here, many client managers and executives have contributed much through their examples as team leaders and members.

Specifically, I want to express my gratitude to

- **ASCE**, for asking and for pursuing this new edition despite my procrastination

- **Carol Hensey**, my partner and spouse, for her patient support in so many ways over the years

- **Ann Somboretz**, for her talented efforts in word processing, formatting, graphics and editing

- **The authors mentioned** throughout this book, for making the effort to share their insights and wisdom

---

The "Hound of Heaven" is also heavily responsible for this work. The Lord of Life has gently but persistently nudged me out of harm's way and into useful service throughout my life.

CONTENTS

FOCUS OF THE BOOK

Work groups and networks are the dominant and pervasive "organizational structure" of our age. Information technology, electronics, hardware, and software have reshaped the workplace. Now more than ever before, people are working on things together.

Groups may no longer always be in the same space or time, but they are more connected in their thinking and their activities than at any time in history.

There are some excellent resources on groups and teams and their development. At the same time, there are many essential and practical aspects of **teams and team development** that are not well addressed or easily available. These include the following:

- Metrics and **measures** for groups and teams

- **Problem-solving** processes for groups

- **Feedback** vehicles and approaches

- Overcoming **separation** in time or space

- Problem **prevention** practices for teams

- **Expectations** for group members

- Getting groups or teams **unstuck**

- **Conflict** resolution and reduction

These and other practical tools for teams are addressed in this new version. In addition, updates of the solid practices presented in the first edition are provided.

CHAPTER 1

TEAMS AND TEAMWORK: THE POTENTIAL

> *Some of today's best-run companies are investing less and less in the talents of individual managers. Instead, they're making major strides by developing management **teams**.*
>
> —Gisele Richardson[1]

Great Potential

Deliberate efforts to develop teamwork in most work groups can clearly improve productivity, communication, schedule maintenance, cost control, work quality, individual satisfaction, and (in the private sector) profitability.

Project teams and task forces need faster team development because they are often transitory, subject to blitz schedules and/or difficult conditions.

Executive committees, boards, and management groups of all kinds would benefit from team development to help deal with their often hard-charging, impatient executive personalities.

Edwards Deming, Tom Peters, and Peter Drucker, to name a few, have heightened our awareness of the need for and value of teams and teamwork. Yet few groups are, in fact, developed to anywhere near their potential!

Why doesn't more work team development actually occur?

---

[1] Gisele Richardson, "Teamwork Magic, Teamwork Secrets," *Boardroom Reports*, 1 December 1988.

The Causes

My experience with management work groups indicates the following most frequent causes of unrealized potential:

☐ Many have come to view team development as an event, like "having a retreat" or "training," rather than as an ongoing process.

☐ Some still equate team development with "good feelings" rather than with tangible bottom-line benefits.

☐ While most of us have an appreciation for teamwork on the basketball court or gridiron, few people can visualize teamwork in the work setting.

☐ Leaders and managers have no practical model or measure for describing teamwork or team dynamics.

☐ Few people realize that the personal chemistry of a group is as important as its professional skills.

☐ Too few leaders know how to improve teamwork, even if they sense that it is needed.

What's Ahead?

This book will address some of these causes through an examination of what teams look like and how they develop, the benefits of team development, the potential critical events in the life of a group or team, several methods for team development, competencies needed by team leaders and members, and other important aspects.

As we get started into the world of "teams and teamwork," here's a nugget to consider. **Following are the bare essentials** for any effective work group, team, or task force—no matter what its field or endeavor:

☐ The **Mission** or **purpose** is commonly known, understood and supported.

☐ **Communication** about the group's issues is open and honest yet also compassionate and sensitive.

☐ **Ground rules**, guidelines, or codes of conduct are known, understood, and supported.

☐ **Leadership** is open to hearing feedback and is responsive in appropriate and nonpunitive ways.

☐ **Members** have a sense of accomplishment and opportunity for improvement as a result of constructive (external) feedback.

CHAPTER 2

WHAT ARE "TEAMS" AND WHERE DO WE FIND THEM?

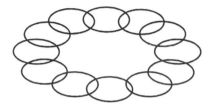

Team Defined

The word *team* may be defined as a group of people, each with different skills and often with different tasks, who work together toward a common project, service, or goal with a meshing of their functions and with mutual support.

Using this definition, most of us are members of many groups that are or could be teams, such as project teams, boards of directors, departments, committees, task groups, offices, and networks!

Some groups are highly interdependent and must mesh their work moment to moment, while others can be more flexible. Typically, most work groups range within the spectrum shown in *Table 2.1.*

Table 2.1. Spectrum of Interdependence in Teams

LOWER INTERDEPENDENCE	HIGHER INTERDEPENDENCE
Consulting firm	Survey crew
Research department	Steel worker crew
Department head council	Blitz project team
Board of directors	Group interview committee
Dental office	Surgical team

Whatever the degree of coordination necessary, some potential for team-work exists. With teamwork may also come increased effectiveness.

Common Characteristics

Whatever the nature of their tasks and whether they work in the laboratory, office, ball field, or on the construction site, effective teams display certain common characteristics. All the following qualities and characteristics are important.

1. The **mission** or **purpose** is known and understood by all group members, who may even have participated in its development. Members support the mission and take direction from it for their work.

    Lack of clear purpose or mission is one of the most frequent difficulties groups encounter. Leaders are often surprised that the group's purpose is not as clear to others as it is to themselves.

2. The group has developed **communication** that is open and direct enough to be able to honestly discuss almost any problem it faces, including its own performance and problems related to performance.

    Lack of open and direct communication is less often a problem of skills than of norms. Norms are set primarily by the organization's traditions and the leader's behavior.

3. **Ground rules** (or a code of conduct) provide guidance for members' actions, practices, and acceptable/desirable team behaviors.

    These are norms set deliberately for group effectiveness rather than by the leader's behaviors or through the organization's traditions—however good they may be.

4. **Leadership** is available in the group, including the designated leader/ manager and others who provide supporting leadership for special tasks such as coaching, quality assurance, marketing, and sales.

    Leadership is a set of skills that usually exist to some degree in many members of a group and become more forthcoming when there is need, opportunity, and encouragement to use them.

5. The group regularly **reviews** how it's doing in several vital areas to avoid "going out of business" through a failure of any of the following:

    ⊛ The **relevance** of its work as to what is really needed by its clients or customers (including internal "customers")

4

⊗ The **quality** of its work as compared with standards and client or customer expectations

⊗ The **impact** of its work and work process on group members, other organizational units, the environment, and other "stake-holders"

Feedback on these three concerns is essential for survival in these fast-changing times!

6. The group has an organization **structure** and member **roles** that are functional, coordinated, and known to all who need to understand them.

Again, leaders are often surprised to learn that members of their group are unclear, confused, or totally in the dark about the organization's set-up or individual roles within the organization.

7. Adequate **resources** must exist either inside or outside the group in order for it to function well. This includes member skills, tools and systems, as well as facilities and budgets.

However, experience has shown the clear superiority of a well-developed team's ability to perform despite shortages when compared with groups that are better resourced but less well developed.

> When a group has these characteristics and features, it is well poised to be an effective team, although it may not yet be one. The overarching feature that pulls all this potentiality into the actuality of a team is synergy.

8. **Synergy** is the breakthrough quality that makes a group greater than the sum of its parts. It rests on the tangible support given to one member by another. Support is given in the form needed at the moment (e.g., time to listen, a pair of hands, encouragement, or a reminder).

Support is the vital aspect of a group-become-team that brings many benefits:

⊗	esprit de corps	⊗	personal joy
⊗	enhanced creativity	⊗	deeper commitment
⊗	collective wisdom	⊗	greater resourcefulness
⊗	stronger productivity	⊗	tougher resilience

These are not the only characteristics of effective teams, but they seem to be the irreducible minimum *(Figure 2.1)*. Other important but more specialized characteristics of effective teams will be offered in later chapters.

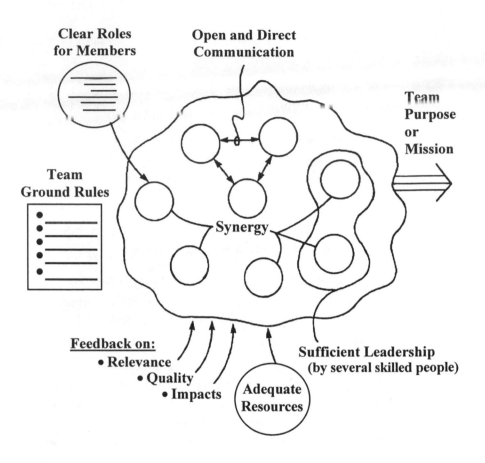

Figure 2.1. Graphic View of an Effective Team

Teamwork Looks Like ...

Over the years, I've asked many managers, leaders, teams, and groups what teamwork looks like when it's happening. In other words, what are the concrete on-the-job actions they associate with teamwork? Here's a short list of what they found:

 Seeking out others' opinions and involvement in matters that concern them before finalizing decisions or plans

- **Being available** when help is needed by colleagues, even if it's inconvenient or requires extra effort

- **Voluntarily offering** your own relevant experiences, ideas, and findings to colleagues who could use them

- **Making your timely contribution** to someone else's action plan or project when requested or needed

- **Acknowledging** a colleague's contribution to a project or sharing credit when working with a client or senior manager

- **Being nondefensive** and receptive to the suggestions, ideas, opinions, and needs of colleagues; making the effort to understand before criticizing

- **Considering the impact** your plans and actions will have on administrative and accounting staff, including allowing time for them to plan their work flow to support you well

- **Being unwilling to criticize** a third party who is absent; not gossiping

- **Coming prepared** to present or participate when you have a role to play in meetings

- **Expressing appreciation** for teamwork extended to you and/or your group members that was helpful

- **Identifying and helping** to pick up loose ends even though they may not be in your area of responsibility; not letting a colleague "twist in the wind"

- **Keeping people advised** of changes, developments, and new information on a task or project

- **Being supportive** of the team's objectives once they are set rather than sabotaging, faultfinding , or being negative behind the scenes

- **Pitching in** when the whole team needs help in meeting a deadline or solving a problem, even if it's "not your job"

- **Trusting the team** to develop consensus on an issue, even if it takes a little more time; hanging in there

---

This list may be just the kind of concrete information your team needs. It could be a great team self-assessment tool.

Notes to Myself about How My Team Measures Up

❀ Mission or Purpose:

❀ Open Communication:

❀ Team Ground Rules:

❀ Sufficient Leadership:

❀ Feedback from Clients:

❀ Feedback from Members:

❀ Structure and Roles:

❀ Adequate Resources:

❀ Team Synergy:

CHAPTER 3

BENEFITS AND VALUE OF A "TEAM" APPROACH

> *Major gains in quality and productivity most often result from teams—a group of people pooling their skills, talents, and knowledge.*
>
> —Peter R. Scholtes[2]

Many still equate teamwork simply with "feeling good" or being "the right thing to do." Yet leaders who have made the effort to develop teamwork in their work groups have noticed some interesting benefits. The following examples were reported by real-life managers from many fields of work, including construction, consulting, public service, education, manufacturing, and information technology.

Benefits Mentioned Again and Again

↗ **Information flow is more effective with teamwork:** Messages get to people and are easily understood. People are more likely to get the same message at the same time and can ask for clarification if they don't understand it. Others also benefit from those questions and answers.

↗ **More people have more knowledge** of problems, projects, contracts, and practices. They know what they need to know for their own work and are better able to contribute appropriately to the work of others.

↗ **Meetings are more goal oriented** and productive, making better use of time. When all team members share the responsibility for using time well, meetings are no longer the sole responsibility of the team leader.

↗ **Better decisions are made** by using a "knowledge bank" of everyone's ideas. The quality of decisions is impacted by having more options available and more perspectives on hand.

---

[2] Peter R. Scholtes, Brian L. Joiner, and Barbara J. Streibel, *The Team Handbook: How to Use Teams to Improve Quality,* 2nd ed. (Madison, WI: Oriel, Inc., 1996; www.orielinc.com).

↗ **Team problems are identified** and brought to the fore where they can be addressed. Better solutions to any problem are more likely by using the collective wisdom of the team.

↗ **Team members learn new skills** by leading and participating in team efforts. The forum created by team meetings provides the best opportunity for learning by actually **seeing** the skills of others as they use them. This is often the only opportunity to see others' work.

↗ **The group becomes more cohesive**, with a stronger sense of belonging. People are more likely to see themselves as a necessary part of the organization or group and are often more committed to it.

↗ **Overall morale improves**; individuals feel better about their own contributions and the group's contributions, effectiveness, and value.

More Tangible Benefits

Although they are clearly beneficial, the benefits discussed thus far can be difficult to measure. In the following examples, leaders claim they can demonstrate that the benefits of teamwork are quantifiable.

↗ A team can **accomplish significantly more** than is possible through individual efforts, regardless of whether those efforts are routine or creative, complex or simple.

↗ **Productivity and time savings** are experienced because tasks are not duplicated. Furthermore, some non-value-adding tasks are more likely to be eliminated or at least simplified.

↗ A team approach **reduces absenteeism and missed deadlines** through a stronger commitment to both the group's members and to its mission or purpose.

↗ Team members are **more willing to be flexible** with one another's changing needs. Team members who are more knowledgeable about and supportive of other members' work, are more willing to adapt when necessary.

↗ Team members have a **vested interest in their organization** and a stronger commitment to its goals. Feeling more a part of the group, they have a higher level of "emotional ownership."

↗ Teams often **set tougher goals and higher expectations** for themselves than their leaders would set for them. Involvement draws on their team aspirations rather than their natural affinity for finding fault with goals they didn't help to create.

↗ Individual team members **feel a greater sense of achievement** and produce higher-quality work. Members feel ownership of both team accomplishments and their own personal contributions.

↗ What is usually seen as **the boss's job is shared,** in turn helping the team in at least two ways: There is an added sense of commitment, and (usually) many heads are better than one!

Quotes from Various Workplaces

↗ "We do our strategic planning and our business planning every year as a management team. This provides us with a better and more practical business plan. More importantly, people are more committed to the plan because it's their plan, they're the authors."

↗ "Division objectives were broken down into smaller segments appropriate to the smaller units. We described the authority of each unit, including dollar budgets available. We then involved a cross-section of technical people from various functions to identify and accomplish the necessary tasks."

↗ "We used a team for coordination and timing of activities for public improvements construction among several city departments, divisions, and outside agencies. The team focused on advertising, purchasing, contracting, preconstruction meetings, job inspection, quality control, and timely payments. It led to the most hassle-free project in our experience."

↗ "We always use a team approach—at least two people—in contract negotiations. Whenever you negotiate, it helps to have more than one person on your side of the table. In addition, one person brings negotiating and contracting expertise, and the other brings technical perspectives."

↗ "We meet as a team of division managers and department directors to jointly decide on priorities for budget proposals for the city manager. This is the only way that works; we've tried everything else and it only resulted in turf protection and budget games."

↗ "Historically, production and sales were at odds, each protecting their own interests. That's not unusual in our industry. Through team-building work efforts, a better attitude has developed between departments. The payoff is that our customer is better served, and we are more competitive."

↗ "Teamwork for site maintenance allowed us to delegate several work areas and move from a centralized to a decentralized crew approach to maintain a 50-acre site complex. We needed no additional staff, morale improved, service improved, and complaints decreased."

↗ "Training programs were developed in a team effort by employee relations staff and line management. The result is more effective training. Another benefit is much-needed improvement in relationships between employee relations staff and manufacturing management."

↗ "Interdisciplinary teams were created consisting of team chief, office engineers, and quality engineers, all working on one project until it was completed. For the first time, these people worked together."

↗ "Our safety specialists don't merely make lists needing corrections for operations. Now, they work with field engineers for safety training and guidance, as a team. There is less faultfinding and more safety."

The "Bottom Line" on Teamwork

Skeptics have asked me over the years whether I can provide documented case examples in which a commitment to teamwork made a big difference to the bottom line. The answer is yes, and the following three examples should suffice to make the point; two of the examples are documented in the available management literature. Significantly, they come from vastly different realms of work—a private, for-profit consulting and design firm; a public, not-for-profit department of transportation; and a multidisciplinary research laboratory of a major university.

↗ **A midwestern consulting and design firm** tripled its net profit before taxes and doubled its size over a 3-year period, primarily as a result of a team approach to leadership and strategic planning by the operations management group, working together with one another and president Fred Weckel. The details are reported in the *Journal of Management in Engineering* (Vol. 5, No. 1, Paper 23102, January 1989; Reston, VA: American Society of Civil Engineers; www.asce.org).

↗ **A large metropolitan district of a state department of transportation** dramatically improved its maintenance productivity over a 3-year period with a 24% improvement in **one** fiscal year. This amazing improvement resulted from participatory planning and management by leaders at all levels, from field foremen to assistant district engineers, led by district engineer Roger Carrier.

Some details are reported in *Output/Input: Department of Transportation Productivity Report* (Vol. 1, No. 4, 1985; Harrisburg, PA: Pennsylvania Department of Transportation; www.dot.state.pa.us).

↗ **The Herrick Laboratories of Purdue University** have long been a nationally known provider of innovations in the multidisciplinary area of heating, ventilating, air conditioning, controls, etc. The research work of the lab has traditionally been supported partly by industry and partly by the faculties of several departments in the College of Engineering.

A multidisciplinary faculty team planning approach, led by the laboratory director Bob Bernhard, resulted in a significant and needed increase in support by both industry and faculty.

Potential Benefits and/or Values of Interest to Me for My Team

↗ More Effective Information Flow:

↗ People Are More Knowledgeable:

13

↗ More Goal-Oriented Meetings:

↗ Better Decisions Are Made:

↗ Team Problems Are Identified:

↗ New Skills for Team Members:

↗ More Cohesive Group:

↗ Overall Morale Improvement:

CHAPTER 4

COMMON MISCONCEPTIONS ABOUT TEAMS

There are some widely held myths and misconceptions that can and do thwart leaders and groups who would like to be part of a team. Because these misconceptions set up false expectations, they make real team development difficult or even impossible.

Let's look at five of the most common misconceptions. Many groups and even some leaders believe that being a team means

- not needing or having a leader
- having a say in all decisions affecting the work
- that certain groups are teams, and others can't be
- just "talking teamwork" will make it happen
- team leadership is more difficult than traditional leadership

No Leadership?

Teams need various kinds of **leadership** for various situations, and they almost always do need leadership. Even "self-managed" teams or groups have leadership, often with several members sharing that responsibility. Leadership isn't a "position" so much as a set of skills or behaviors that many people in a group may have. Leadership within a team may be relatively autocratic, democratic, or a mixture of both.

It's important to remember that teamwork is only partly defined as an effective relationship between leader and member. Much more so, it is an effective relationship between member and member. That is where teamwork really pays off.

15

Having a Say?

Planning, problem solving, and decision making in groups may be

⚽ **leader-centered**: the leader plans, decides, and advises the team

⚽ **group-centered**: the entire team (including the leader) discusses the plans **together**, within any necessary constraints

⚽ **both-centered**: the leader welcomes input and suggestions before making his or her plans/decisions

All three modes *(Figure 4.1)* can be appropriate and effective, depending on the situation. The more the team's leader needs team input and/or commitment, the more important it is to **use the group!**

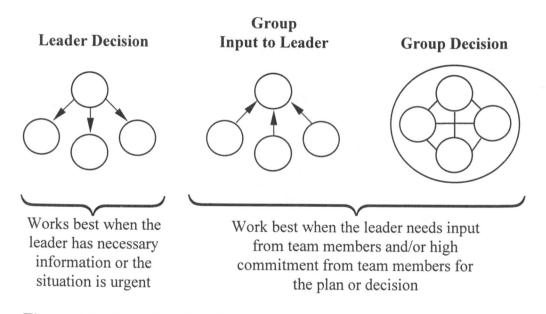

Leader Decision

Group Input to Leader

Group Decision

Works best when the leader has necessary information or the situation is urgent

Work best when the leader needs input from team members and/or high commitment from team members for the plan or decision

Figure 4.1. Planning, Problem-solving, and Decision-making Modes

Tannenbaum and Schmidt present an elegant model of this concept in the *Harvard Business Review* classic "How to Choose a Leadership Pattern" (May-June, 1973, Reprint #73311).

The wise leader does not dwell at one extreme or the other but chooses how to involve team members as appropriate to the issue at hand. Examples in which team members ought to be involved include the following:

⚽ Training new staff and one another

⚽ Setting team goals, targets, and objectives for the team

16

- ⊕ Reviewing team performance toward goals
- ⊕ Discussing problems or priorities in which the leader does **not** have all the facts
- ⊕ Reviewing any issue for which it's very important to have team commitment

Who Can Be a Team?

Groups, teams, and potential teams exist in many places in all organizations, not just in the lines, boxes, and departments of the typical organization chart. Many potential teams are in departments. However, other potential and actual teams include

- ⊕ committees and task groups
- ⊕ informal and ad hoc groups
- ⊕ networks and service lines

In fact, groups that have the greatest potential for becoming more effective through increased teamwork are those in which the members

- ⊕ have a common purpose or focus
- ⊕ are interdependent with one another
- ⊕ face tasks and problems beyond their own resources
- ⊕ have a means of communication
- ⊕ are not inherently competitive with one another

This includes many more potential teams than most folks imagine. It certainly includes traditional hierarchical departments, matrix organizations, project teams, committees, and boards.

Sometimes I ask people to identify how many groups, teams, committees and task forces they participate in at work. Many folks come up with 10 to 15!

"Talking" Teamwork

Probably the **greatest mistake** leaders can make is to talk about teamwork without facilitating or modeling it themselves. Although this is sometimes understandable, it can create cynicism among group members and make the reality of teamwork even less likely.

The most frequent cause of this problem is the need for "control." Managers have a good deal of training in maintaining control and in many cases have been rewarded for being "in control." Yet team leadership requires, by its very essence, the loss of (some) control—to the team of course.

17

Leaders owe themselves, the group, and the organization they serve a knowledgeable experience of team leadership and teamwork. Otherwise, the traditional hierarchical approach is probably a better deal for everyone involved.

Team Management is "Difficult?"

Finally, many assume that managing teams is more difficult than managing groups in the "traditional" manner. Not necessarily so! In most cases, managing a team is easier on the nerves and the spirit, although not necessarily less time consuming.

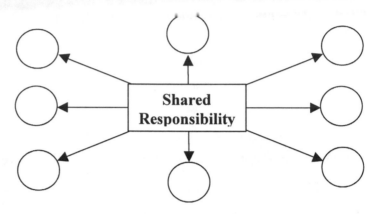

My own learning came in my last two management assignments in the Engineering Division of P&G, a great industrial firm and an American institution for over 150 years:

⚽ As a young section manager for three older and wiser civil engineering group leaders, I truly stumbled into this discovery. Because I was unsure what all my new responsibilities were, these experienced and professional colleagues were very helpful to me and to one another. **Together**, we got the management job done.

⚽ Responding to an internal "job bidding" opening, I applied for the position as manager of a personnel section. I was trying desperately to learn new and unfamiliar skills related to salary administration, hiring and firing, training and development, affirmative action, and temporary staffing. It would have taken me much longer to learn (less well) without the help of the professional women and men there.

> In both situations, it became clear to me that we made the most progress as a team when I learned to ask for others' ideas, thoughts, and assistance of all kinds. Help was never slow in coming. It also made the job more satisfying, as well as easier to learn and do!

What Are Some of My Own Beliefs About ...

  Teams:

  Teamwork:

  Leadership:

🕎 Decision Making:

🕎 Who Can/Can't Be a Team:

CHAPTER 5

LEADER AND TEAM MEMBER ROLES

> *No sharp distinction can be made between leadership and membership functions, between leader and member roles.*
>
> —Ken Benne and Paul Sheats[3]

Leader/Team Roles

Groups may consist of people who have similar or very different roles or tasks. A group may not be required to work as a coordinated or integrated unit, such as a group of dentists, attorneys, technicians, or truck drivers. In such cases, the roles of group members may relate mostly to their own individual jobs, functions, or objectives.

By their very nature, teams require additional roles of their members. These additional roles relate directly to their team responsibilities.

Team members have an obligation (and an opportunity) to support one another and function in an interdependent way. This suggests additional and different roles than those needed to work effectively as an individual.

In addition, team leaders have some additional roles that go beyond "traditional" management tasks *(Table 5.1)*. Note how many tasks or roles apply to **both** leaders and members. As my friend and colleague Ken Gibble of Gibble, Norden & Champion (Old Saybrook, CT) said "I get it! Team members can't just be passive participants!"

The table is adapted from a similar list used in training adult and boy leaders in the "patrol method" used by the Boy Scouts of America[4].

---

[3] Ken Benne and Paul Sheats, "Functional Roles of Group Members," in *Group Development*, 2nd ed., ed Leland P. Bradford (San Francisco: Jossey-Bass, Inc., 1978; www.josseybass.com).

[4] www.bsa.scouting.org.

Scouting offers greater insight into teams and teamwork as well as leadership and training than does much of the available management literature!

Table 5.1. Roles (or Tasks) of Team Leadership and Membership

ROLE (OR TASK) (1)	TEAM LEADER (2)	TEAM MEMBER (3)
1. Knowing individual members and their skills and resources	Yes	Yes
2. Understanding the purpose and needs of the team	Yes	Yes
3. Getting and giving information when needed	Yes	Yes
4. Planning, scheduling, organizing, and goal setting as a team	Yes	Assisting
5. Using the team as a resource; asking for advice, help, and support	Yes	Yes
6. Leading and guiding the team in its work or some part of it	Yes	As needed
7. Representing the team (or a member) to others	Yes	Yes
8. Coaching individual team members on their needs	Yes	When able
9. Setting an example for others in- and outside the team	Yes	Yes
10. Producing results as a team and as a member	Yes	Yes
11. Evaluating team performance, needs, progress, and problems	Yes	Assisting

In this list of roles, we can see some of the most fundamental differences between teams and traditionally managed work groups. The **interdependence** of leader and team and of member and member is quite clear.

Note how many of the roles (or tasks) of team leaders are also required, to some degree, of team members. Members capable of fulfilling roles in planning, organizing, guiding, and coaching will have an opportunity to contribute there.

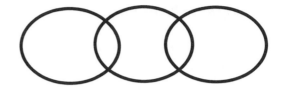

Some Real Applications

Project Team Leadership

Within the same engineering project team in which **Ralph** is project manager (team leader), **Mary**, an experienced engineer, serves as coach for several younger staff members. **John** represents the team at operations committee meetings; and **Carl**, a young recent college graduate, serves as a technical resource on software for the team.

Project Team Leadership

Ralph, project manager, is team leader; his duties include making assignments and reviewing work, as usual.

Mary, senior engineer, coaches several younger staff engineers in their work assignments and answers their questions.

John, senior engineer, represents the whole project team at the weekly meetings of the operations committee, making reports and gathering needed information.

Carl, a young engineering graduate, provides hands-on help with software to all in the group who want it.

All four are in leadership roles supporting their project team.

At this point, a traditional manager, who has learned to always be in control, may be uneasy about how Ralph has shared power and what is usually thought of as a manager's prerogative (or duty). Some may think Ralph is weak, hiding behind the team, perhaps unable or unwilling to make tough decisions. However, the truth is that it takes more courage and self-confidence to do what Ralph is doing. He is encouraging his team to contribute all they are able to wherever they can, and he's done it based on their abilities and probably based on their interests as well.

Because people will be contributing to their fullest in both abilities and energy, Ralph's team will have higher morale, productivity and effectiveness than they would otherwise.

Executive Team Leadership

Within a medium-sized, multioffice firm in the business of consulting engineering and design, **Frank** is the chief executive officer (CEO). While he has all the usual responsibilities that go with the job, Frank is very comfortable asking **Carol**, his executive secretary, to handle many matters directly with the vice presidents in his absence.

Frank also asks **David**, vice president of marketing, to sit in as chief negotiator of all contracts over $200,000, something Frank has responsibility for. In both these cases, Frank relies on his coworkers beyond their normal duties to the executive team.

Executive Team Leadership

Frank, CEO, is president of the firm, leader of the executive team, and responsible to the board.

Carol, executive secretary, handles many matters for Frank directly with the vice presidents in his absence during trips.

David, vice president of marketing, serves the firm as chief negotiator for all contracts over $200,000 in fees.

Again, this sharing of leadership power and responsibility on behalf of the executive team is warranted on the basis of team member abilities and interests.

- ✓ **Carol** is able to keep the executive team connected and smoothly functioning during Frank's frequent business trips.

- ✓ **David** is able to make major commitments for the firm (as others do) but with an unusual flair for excellent deals that improve the firm's earnings.

This sharing of power and assisting with roles, tasks, and competencies among members and leader will also be seen in the following chapter on stages of group development.

CHAPTER 6

STAGES OF GROUP DEVELOPMENT

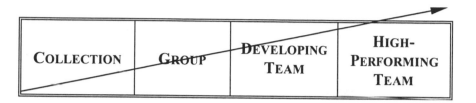

Most groups have a natural potential for **growth and development** over time toward becoming high-performing teams. Tuckman, Francis and Young and others have noted that groups may progress through **stages**.

Stage I (COLLECTION)

All potential teams start as a collection of people. They may come together quickly (e.g., at the start of a project), or they may gather slowly over time (e.g., a growing branch office). Teams may even come about through the merger of two groups.

In the beginning, this collection of people will tend to be polite, somewhat guarded, and interested in learning about other members. There may be several "beginnings" If people are added to the group over time. The key goal at this stage is to move beyond "polite talk" to openness.

Stage II (GROUP)

As people discover more about one another, including their personalities, values, skills, and interests, they begin to open up and to be more candid with one another. They may also "open up" on the leader (with criticism).

This opening up will usually result in sparks of conflict and confrontation as well as feelings of frustration and difficulty. Subgroups or pairings may form. People have begun to identify with the group but not necessarily with one another.

Stage III (Developing Team)

When positive team leadership is present (regardless of whether there is one or several leaders), the group **may** move on to the next stage. The potential advantages of being a team may emerge. People may begin to see others as resources and the group as a useful vehicle for working effectively.

During this time, members will begin to establish norms, goals, roles, practices, and even procedures within the group, either formally or informally. Decision making and problem solving may be stressful or laborious, but they are usually more successful than in stage II. In addition, the work will feel more satisfying to team members.

Stage IV (High-Performing Team)

Able to work out problems, goals, issues, and difficulties, the developing team is poised to move forward to high performance. Yet few will make it, perhaps because of the necessary give and take, the potential for personal closeness, or the higher personal accountability this calls for. The few teams that reach the high-performing stage enjoy a number of strengths and benefits.

Among the strengths and benefits are effortless synergy, high morale, and excellent productivity. Although "conflict" is frequent, it doesn't look like conflict. It looks more like spirited problem solving. Differences are valued. Loyalty to the team is strong, and caring among members is evident.

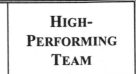

HIGH-
PERFORMING
TEAM

Overall

The four stages of group development can be summarized as collection (stage I), group (stage II), developing team (stage III), and high-performing team (stage IV). Tuckman and others have summarized these four stages as …

FORMING, STORMING, NORMING, AND PERFORMING.

Task groups, committees, and work groups of all kinds may go through several stages of development. It pays to know what the stages are, what to expect, and how to move on if you want to!

Table 6.1. Stages of Group Development

STAGE I COLLECTION	STAGE II GROUP	STAGE III DEVELOPING TEAM	STAGE IV HIGH-PERFORMING TEAM
People are • cautious • guarded • wondering There is little visible disagreement. The collection lacks an identity. There is little investment in the group function. People are watching for the norms to see what is okay or expected of them.	The **group** is developing • identity • purpose • interest People are taking risks and getting to know one another. Conflict is in nonproductive fits and starts. Levels of frustration and/or confusion are high. People develop pairs and cliques.	**The emerging team** is developing • goals • roles • relationships Members are learning to appreciate their differences. Conflict is usually about issues, not egos. Communication is open and clear. There is a sense of belonging. There is a sense of progress. The team is enjoying its work.	**The team** is acting on common goals with • synergy • high morale • high productivity Shifting of roles from one to another is easy. Differences are valued. Members look out for one another's interests. Efforts are spontaneous and collaborative. Members share all relevant information. Conflict is frequent and often looks like problem solving.

(0)   **(3)**   **(6)**   **(9)**   **(12)**

Most of the insight into the development of this table, which is based on our consulting experiences, was provided by Dave Francis and Don Young in their book *Improving Work Groups: A Practical Manual for Team Building* (San Francisco: Jossey-Bass, Inc., 1992; www.josseybass.com).

27

Assessing Stages

In their practical book, *Improving Work Groups*, Francis and Young show the four stages of group development in the form of a clock. *Table 6.1* makes use of this notion by showing the numbers 0, 3, 6, 9, and 12, which enable a group to more easily identify its current stage of development.

> In trying to assess the group's current stage of development, it helps to look over the **descriptions in each stage as a whole** rather than at each item individually.

For example, a group of five senior managers who together make up the operations management team rated their team individually, with scores of 7, 7, 7.5, 6, and 8. So the average rating of their team was about 7, with a range of 6 to 8 (or part way into stage III). As they discussed their individual ratings, and more importantly, their **perceptions**, the group set its collective goal to be at stage IV (a rating between 9 and 12) within a year. They also identified some of the things they needed to do to get there.

It may be useful to note that over many years in our practice, as management teams have assessed **themselves**, their ratings form a normal distribution. We also noted the following:

- Most team's self-ratings average about a 6 on the 0 to 12 scale.

- Team members tend to be quite honest in their ratings.

- Members' ratings are often comparable to one another—within 2 points on the scale (higher or lower).

- This simple tool makes it easy for members to talk about teamwork.

As team members' ratings are collected, it is helpful if the **team leader goes last** because his or her ratings may be lower or higher than most others, and he or she may inhibit members whose ratings differ.

How is This Helpful?

A real working group can easily evaluate its current stage of development. More importantly, they can discuss practical ways to move toward greater levels of teamwork, effectiveness, and satisfaction, if they choose to.

There are several simple steps leaders can take with their team members:

☞ Ask team members to **evaluate** their stage of group development (using *Table 6.1* as a guide).

☞ **Plot** their numeric responses on a flip chart or white board, as seen here:

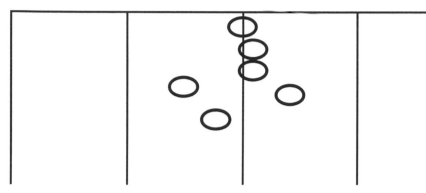

☞ Ask the group: "What **practical and inexpensive** things can we do, as a team, to further develop?"

☞ **Write** those ideas and suggestions on the flip chart.

☞ Ask the group: "Out of all these good individual ideas, which ones do we want to **commit** to do?"

☞ Lead the discussion and **summarize** the group's commitments.

This plot of responses can serve as a **benchmark** for checking progress in the team's development. Looking at the Stages of Group Development Chart periodically can do a lot for your team!

CHAPTER 7

HOW TO DEVELOP FURTHER AS A TEAM

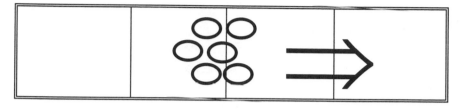

Once they understand that things can be different, many teams and leaders decide they'd like to see what's involved in developing further as a team. The drive for this may be a sense of great opportunity or simply dissatisfaction with things as they are.

Dissatisfaction may stem from noting that group morale, productivity, mutual support, and/or communication are just **not** what they ought to be. It's important to remember that the world seems to work against teamwork, with shortages, changes, staff turnover, and overload.

What To Do?

When leaders or teams sense it's time to do something about their level of teamwork, questions like the following may come up:

- O Do we need a retreat?
- O Do we need some training?
- O Will it be expensive?
- O Will anything really make a difference?
- O Is there a certain procedure or formula?
- O Are there any experts on this?

Good News

The answers to these questions have some really good news: Team development can usually be very functional and get some otherwise needed tasks done for the group!

There is usually little need for training, although there is often much need for **learning**. Most real learning is learning by **doing** or learning from **others** as the group works at necessary and useful tasks.

In team development, it is important to determine what to do next. First, there is no one right way to do this because each group is unique. What needs to be done can only be assessed from **within the group**, where the data exist in the form of group members' perceptions, feelings, needs, concerns, hopes, goals, and opinions.

This takes the load off the leader because only the group knows what it needs next for its own development. The leader's role then is to **facilitate the group** in discussing this matter to consensus and elicit the collective wisdom of the group.

Typical Team Development Tasks

Here are some typical tasks that can assist in team development:

O Define the group's **mission or purpose**, or refine it if it already exists. The mission should be the main objective, the reason for the team's existence, its calling.

O Review the team's **performance** against its mission: How are you doing in fulfilling your mission or purpose? This is a way of giving the mission more meaning.

O Define the group's **vision**; that is, where do you want to be in the future (e.g., 3 to 5 years from now)?

O For a project group or task team, set specific project **goals** as an essential early step.

O Discuss the group's work **norms and traditions**. Specifically, how do you want to work together in the future?

O Assess, together, the group's **current stage** of development and then identify barriers to further development as a team.

O Prioritize these **barriers** to further development, or discuss which (if any) to tackle.

O As a group, carefully define the various **roles** of group members. If definitions of the roles already exist but are incomplete or out of date, refining them is usually valuable.

○ Create some **needed plans** for the group, such as a strategic plan, a quality improvement plan, a list of annual objectives, a marketing/sales plan, a budget to support plans, a cost-reduction plan, or a human resources plan.

○ Evaluate in detail the **group's performance** against some meaningful plan, such as last year's objectives or budget. It's important to celebrate successes as well as to address problem areas.

○ Design a process for obtaining **useful and specific feedback** from clients/customers, sister groups, or both. Then make use of the feedback to improve services and client satisfaction.

○ Participate in an advanced **team skill learning event**, perhaps one about creative conflict management, understanding personality types, or facilitating group meetings. Learning as a team multiplies the value of the education.

○ As a team, participate in a challenging and **enjoyable experience** such as a canoe or bike trip, whitewater rafting, or another outdoor challenge. Be sure to choose knowledgeable leadership!

○ Refer to *Figure 2.1* and evaluate your group against the common characteristics of an **effective team**.

○ As a group, identify ways to **improve** the team's current performance on each of the characteristics. Brainstorm; seek all ideas without criticizing.

○ Set a date for 1 to 3 months ahead when the group will meet to review **specific progress** on each of the ideas implemented.

Reinforcement Approach

Take a reinforcement approach to team development. This approach recognizes the following:

○ Even in the worst of teams, members occasionally show helpful teamwork behaviors.

○ Leaders get more of what they reward, particularly over the long haul.

○ Lack of teamwork is often a leadership problem, not a people problem.

For the reinforcement approach, leaders must be sufficiently desperate and therefore disciplined to be able to acknowledge or recognize teamwork behaviors at least 20% to 25% of the time.

Leaders must also avoid criticisms (of team members) that are not offered in a helpful way or in a calm and professional manner.

The results of this approach can sometimes be dramatic. Few leaders intentionally sabotage teamwork in their groups, but many are so focused on problems that **they fail to notice what's working well**.

Positive Expectations

Only in recent years have a few leaders become aware of the awesome potential of positive expectations on individual and/or group performance and development.

Field research in the laboratory, office, and real-time workplace clearly confirms the uplifting power of sincere and positive expectations and belief in others, individually or collectively, well beyond past performances.

Most of us have heard the old cautious (and pessimistic) expression: "I'll believe it when I see it!" For managers and leaders, someone has said it would be more appropriate to say, "I'll see it when I believe it!"

This well-turned reversal is more than cute; it's essential, and it underlines the essence of a positive expectation: It has to be more than words; it must be believed in to be believable. Only when something becomes believable to the team will it have the potential positive power to become reality through performance and results.

Positive Humor

Almost as underappreciated as positive expectations is positive humor. Gentle, natural, and self-effacing humor can provide an important team lubricant. Humor facilitates candor and straight talk, mollifies disagreements and conflict, and alleviates potentially embarrassing situations or situations in which some team members may be taking themselves or their positions much too seriously.

Humor in this sense isn't about joking or witticism, and it certainly is not about clever putdowns at someone else's expense. It is about the naturally funny things that pervade most of life even in otherwise serious moments.

This facilitative team-developing skill begins with the ability to laugh at oneself—to find the naturally funny part of life as it happens. As for myself, I find there is no shortage of dumb and unintentionally funny things I do or say.

34

From there, it is easy to help others in the team find humor in themselves, institutions, the team, and the organization. But humor begins with modeling it in **oneself**.

Goals and Goal Setting

There are some strong proponents for goals or objectives for individuals and groups in most organizations. Even bureaucracies can be very goal oriented.

Some people believe goals must flow from the top of the organization; others feel keenly that goals must be set from the bottom up. Both are probably right in most situations. That is to say, those at the top have important broad perspectives, and those at the bottom have important practical knowledge. **Both** elements are needed in effective goals or goal setting.

When people ask where goals ought to come from, I'm likely to say, "from the **real bosses: your clients, customers, users, beneficiaries**. It's the folks you're trying to serve that ought to have the most input into your goals." Of course, that's not totally true, but it is important.

The best input for team goal setting seems to include feedback from clients on quality and relevance; priorities communicated by management; and the team's own views, hopes, and concerns.

Most teams want to serve the organization and its clients better than they are sometimes permitted. Given the opportunity to do meaningful goal setting, they will usually set more challenging goals than the organization would dare set for them!

The Ultimate Recipe (and Acid Test)

A number of management teams have noticed that they developed fastest and/or best when they had a difficult or challenging time and succeeded in overcoming it together.

The three elements are **sharing, struggle, and success**. When success of some kind is an outcome, the more struggle there is and greater is the sense of team that emerges from it.

Trust is often mentioned as an outcome of shared struggle. Any experience that tests a group may also allow its members to discover that they can indeed trust one another. Increased trust leads to stronger team cohesion and performance.

The practical value of this phenomenon is that team leaders can use difficulties that occur naturally in life and work. These difficulties can bring a team further along if they see the difficulties as challenges to overcome together rather than as awful problems to complain about.

Using struggles to further develop a team may well be the acid test for team leadership. Leaders must lead their teams into the struggle, straight in the face of team members' pessimism, which may seem to be justified and/or pervasive. A number of memorable sports teams have demonstrated how this works.

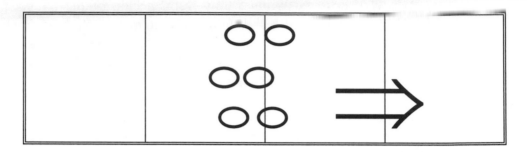

These struggles may occur internally (within the team) or externally (imposed from outside the team); both will be useful for team development. The successful **leadership recipe** is as follows:

○ **Believe** in the team's ability.

○ Let that belief **show** in what you say.

○ Take the **lead** in what needs doing.

○ Allow others to feel pessimistic as long as their **behavior** works toward overcoming the struggle.

Other Means of Team Development

Other means of further developing teams will be covered in the chapters ahead. Other excellent resources for team development include the following:

○ *Improving Work Groups: A Practical Manual for Team Building*, by Dave Francis and Don Young, published by Jossey-Bass, Inc., San Francisco, 1992; www.josseybass.com.

○ *The Team Building Tool Kit: Tips, Tactics & Rules for Effective Workplace Teams*, by Deborah Harrington-Mackin, published by AMACOM, New York, 1993; www.amanet.org/books.

CHAPTER 8

MOST IMPORTANT TEAM TOOLS

One of the major reasons some teams do well and other groups struggle is that effective teams learn to use, and actually do use, a few tools that improve their problem-solving skills and reduce frustrations.

Groups needn't become skilled at all the quality management tools available. A **few simple tools** can make the critical difference:

- ✗ Team charter
- ✗ Team ground rules
- ✗ Typical agenda
- ✗ Problem-solving process
- ✗ Flowcharting process
- ✗ Conflict resolution process

However, the other necessity is a leader with an **awareness of process** and a **facilitative style and skills** to help the group select and use these tools.

Team Charter

One of the most important contributions of "partnering" in construction is the notion that teams need a **charter.** At a minimum, the team charter needs to include the following:

- ❌ Team **purpose** or mission
- ❌ Specific team **goals** as appropriate
- ❌ Time, place, and frequency of **meetings**
- ❌ Means of **staying in touch** between meetings

These charter elements help the group focus on its common **work objectives** and provide a means of basic communication.

Team Ground Rules

Our practice suggests that effective teams in all walks of life and work share some ground rules or guidelines for team behavior. They may be written or not, but group members know what they are and consider them important!

New teams or old teams facing new situations have a great opportunity to create their own ground rules. Following is an excellent sample set of **ground rules** created by a Washington, DC–based research and development group:

1. We **speak candidly** within the team.

2. We **respect** other members and their jobs as valuable.

3. We are **honest** about deadlines, and we honor them; we say "no" when appropriate.

4. We offer praise, encouragement, and **positive reinforcement**.

5. We are **accepting of criticism** and constructive when offering it.

6. We **don't badmouth** team members or others.

7. We strive for **clear communications** both in sending and receiving.

8. Regarding communication on **sensitive matters**:
   - ❌ We are aware of our tone.
   - ❌ We use email carefully.
   - ❌ We take our concerns to those involved, not to a third party.

9. We are explicit about team goals and objectives and take **personal responsibility** for them.

Note how **specific and practical** these team ground rules are. They helped the team improve almost 2 points on a 0 to 5 point scale over 15 months!

Typical Agendas

Whether groups will be short-lived (e.g., project teams) or long-lived (e.g., an operations group), they will benefit from using a standard or pro-forma **agenda**. Here's a sample:

- ❑ Quick **review** of the agenda, fine-tuning it for this meeting
- ❑ Member **updates** for the team on progress and problems
- ❑ **Reports** on studies assigned at a previous meeting
- ❑ Matters requiring discussion and **decisions**
- ❑ **Assignments** as/if appropriate, including:
  - ✓ the charge (task)
  - ✓ who all is to do it
  - ✓ when it is expected

> **Note:** Some groups call this their **To-Do List**, and it is a crucial activity for effective groups.

For team meetings that went unusually well or badly, we recommend that this be the final agenda item:

- ❑ Quick **critique** of the meeting, in which the group identifies positives and needed changes on a flip chart or whiteboard *(Table 8.1)*

Table 8.1. Sample Meeting Critique

POSITIVES	CHANGES NEEDED
Things that contributed positively: ✓ Members were on time. ✓ Joan was well prepared to present study. ✓ Tom helped us get back on track.	Negative things that we need to do differently: ✗ We didn't stick to our agenda well. ✗ Tangents wasted a lot of time. ✗ We didn't stick to our ground rules.

Problem-Solving Process(es)

Many groups of highly intelligent and wise professionals or executives are prone to get tangled up, confused, and (finally) frustrated if they don't follow some simple process for handling challenging or complex problems.

When they **don't** follow a simple process, the result is usually the **CRAPOLA process** (circular, repetitive, argumentative, personal, opinionated, and leading anywhere).

The process I like **best for helping groups** handle typical problem situations is somewhat simpler than most you'll find in the quality management references. I call it the **STOP process**:

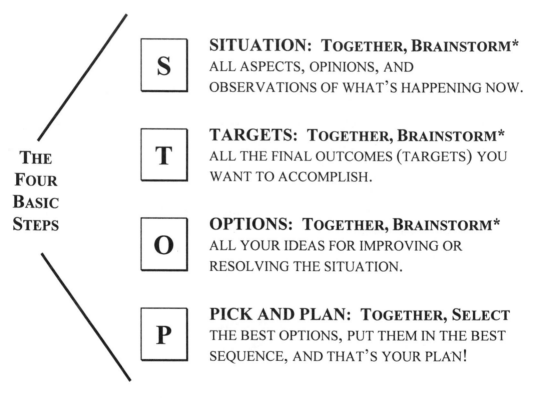

THE FOUR BASIC STEPS

S SITUATION: TOGETHER, BRAINSTORM* ALL ASPECTS, OPINIONS, AND OBSERVATIONS OF WHAT'S HAPPENING NOW.

T TARGETS: TOGETHER, BRAINSTORM* ALL THE FINAL OUTCOMES (TARGETS) YOU WANT TO ACCOMPLISH.

O OPTIONS: TOGETHER, BRAINSTORM* ALL YOUR IDEAS FOR IMPROVING OR RESOLVING THE SITUATION.

P PICK AND PLAN: TOGETHER, SELECT THE BEST OPTIONS, PUT THEM IN THE BEST SEQUENCE, AND THAT'S YOUR PLAN!

*Remember: Brainstorming takes **ALL** contributions WITH**OUT** criticism.

Figure 8.1. The STOP Process

It's very helpful to use a **flip chart** for the STOP process (use at least one sheet for each step). That allows the group to jump back and forth between steps—which they will do!

Flowcharting Process

People often make flowcharting too complex and rule-bound for many everyday applications for which it can be quite helpful. Further, when groups try to prepare flowcharts, the process needs to be …

SIMPLE, VISIBLE TO ALL, and EASY TO CHANGE!

When it meets these three criteria, groups can and should flowchart lots of things, including

- ✂ a meeting plan
- ✂ a simple project
- ✂ phases of a larger project
- ✂ any work-related process

Figure 8.2 presents an example from an actual team of research professionals from Cleveland who focused on improving their weekly meetings:

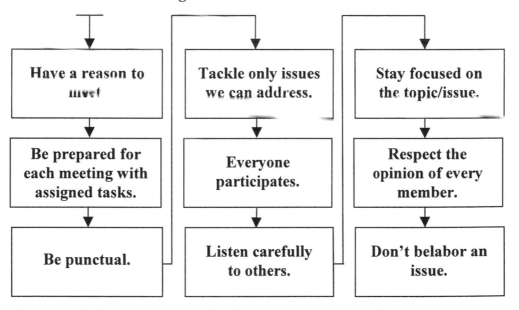

PROCEDURE: Create most of the likely activities (on self-stick notes or index cards). Place them in a logical "flow" of time. See if you need to move, add, or change any notes or cards. Add dates and responsible personnel as appropriate.

EXAMPLE: Our Meetings

Have a reason to meet	Tackle only issues we can address.	Stay focused on the topic/issue.
Be prepared for each meeting with assigned tasks.	Everyone participates.	Respect the opinion of every member.
Be punctual.	Listen carefully to others.	Don't belabor an issue.

Figure 8.2. Flowcharting with Self-stick Notes or Index Cards

Conflict Resolution Process

When groups get trapped in nonproductive or nasty conflicts, it's usually because of one or more of the following reasons:

⚒ Someone feels attacked (whether or not they were).

⚒ The group rushes to find solutions before feelings are sufficiently vented.

⚒ One member believes he or she has the answer and pushes it.

⚒ Several members compete, each with his or her own answer.

At times like this, special ground rules called **conflict management guides** are necessary. My colleague John Parmater and I developed the following guides for tough conflict situations:

1. Begin with the **goals** in mind.

    ⚒ Discuss the goals of the project and the meeting.

2. Put **relationships** ahead of tasks.

    ⚒ Maintain rapport above all.

3. Take **responsibility** for communication.

    ⚒ Know what you want to communicate.

    ⚒ Know whether you're achieving that.

    ⚒ If you're not, do anything else.

4. Understand and compensate for others' …

Social styles:	**Conflict styles:**
⚒ Analytical	⚒ Friendly helper
⚒ Driver	⚒ Sturdy battler
⚒ Expressive	⚒ Compromise seeker
⚒ Amiable	⚒ Conflict avoider
	⚒ Problem solver

Keep in mind that it's still important to follow a simple process here as well. Maybe more important!

Which Team Tools Could Most Help My Team?

⚒ Team charter (e.g., mission, goals, etc.)

⚒ Team ground rules

⚒ Typical agenda

⚒ Problem-solving process

⚒ Flowcharting process

⚒ Conflict resolution process

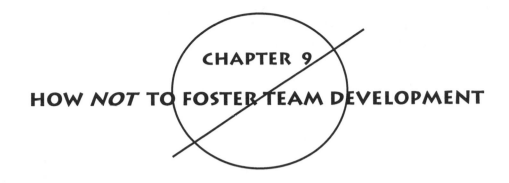

CHAPTER 9

HOW *NOT* TO FOSTER TEAM DEVELOPMENT

> *Scott Adams has become an American organizational icon with his friends Dilbert™ and Dogbert™.[5] These cartoon characters too often reflect reality! For example, a client CEO who'd been sent several Dilbert™ books ignored all of his many clues until he was replaced—ahead of schedule!*

You're Joking, Right?

I almost didn't write this chapter. But every week brings to light new situations in which executives, leaders, groups, and members, sometimes well intended, actually **hinder team spirit** or teamwork. In addition to doing things to facilitate teamwork, it is also necessary to avoid doing things that lessen the teamwork of a group.

In many cases, people are unaware of their team-destructive actions or words that undermine their hopes for having an effective high-performing team. Some of this sabotage is subtle; some is obvious, at least to others. In many cases, these problems continue because of a lack of honest feedback to team members, particularly to leaders from their staff.

Overtalking

Some leaders simply talk too much as they meet with their groups. They may overtalk for many reasons, such as wanting their people to be well informed, overexplaining or justifying management decisions or actions, expecting themselves to have all of the answers, not being able or willing to really listen to people, and not wanting to lose control of a meeting or conversation.

---

[5] Dilbert™ and Dogbert™, United Media, Inc.

Researchers tell us that listening is the most difficult aspect of communication because we must set aside (for the moment) our own opinions, beliefs, and points of view. However, listening is also the most powerful and useful part of communication, simply because we become well informed. Every team needs and appreciates a leader who takes time to listen and is well informed.

Ego Flatulence

Much of the conversation we hear has as its central theme, "Aren't I great?" or "Look what I've done." This strong human tendency is apparently a natural need or drive. I call this ego flatulence (among other things). Although some of us are very clever at disguising our ego flatulence, it can still often be spotted with a little practice.

Closely related to ego flatulence is **CRAPOLA**, which goes beyond bragging, self-promotion, or exaggeration to actual fabrication of one's knowledge and accomplishments.

Ernest Hemingway thought it was likely that we each have what he called a "built-in shockproof crap detector." A necessary tool for living and working and relating to people, this "crap detector" enables most of us to spot ego flatulence or CRAPOLA easily.

Leaders and managers, especially, carry an unfair burden in order to be respected by their staffs. They are expected to have many skills, no noticeable faults, and to be humble about it all. Few meet this expectation. However, managing one's own ego flatulence can be learned; it requires more discipline than skill, but the payoff is that one is far more likely to be respected and appreciated.

Power Plays

Power plays are verbal comments designed to win points or arguments. They are manipulative in nature. Consultants Theo Wells and Tom Parker developed several delightful examples of typical power plays in cartoon form in 1976. Here are some of my favorites:

Ø **"Be specific; give me an example."** (*Hidden message:* "I can shoot down your example and prove you wrong overall.")

Ø **"Now let's be fair!"** (*Hidden message:* "You're not following my rules.")

Ø **"Can you prove that?"** (*Hidden message:* "I know you can't or I wouldn't have asked.")

You can no doubt think of some of your own, or perhaps you recall some power plays that have been used on you. Actually, each of these comments can, under many circumstances, be legitimate and appropriate. They **become power plays** when they are used for a hidden objective in order to manipulate another person.

Some people learn these (and other) forms of manipulation very early in their lives and aren't aware that they use them. Because they arouse suspicion, confusion, and irritation, power plays are best avoided!

Now Hear This!

Overuse of authoritarian or directive communications can also get in the way of team development. It can cause passivity and over-reliance on the leader or manager. It can cause foot-dragging performance, complaining and blaming, and general avoidance of responsibility. It can lead to "malicious obedience," in which people obey directions they know to be wrong.

By no means is directive communication to be avoided altogether. When it is inappropriately or extensively used however, problems can occur and teamwork will suffer.

Public Criticism and Putdowns

The old management maxim, "Praise in public and criticize in private," is generally good advice. Yet there are some very hard-driving group managers and members who fail to follow it.

Hardly anything is more destructive of morale and spirit than public criticism, sarcasm, or ridicule. The more subtle forms are also destructive, sometimes more so. Occasionally, there may be some unexpected team development: The group may unite against a common enemy, the sharpshooter himself!

One particularly vicious kind of public criticism sometimes practiced by group members or managers is in the form of hard-hitting and pointed humor, commonly called "putdowns." Putdowns are so vicious because they are almost immune to response.

The recipient is expected to take it like a "good sport" and show that he or she has a sense of humor. What usually happens instead is that the recipient is left to hurt inside **or** show that he or she is hurt by responding. Either way, the individuals and the group will lose, and teamwork suffers a major blow.

47

Preoccupied with the Pyramid

Leaders and executives sometimes behave in ways that leave the team clear about who is important or who thinks he or she is on the way to the top of the hierarchy or organizational pyramid. Several quality and service researchers, including Edwards Deming and Karl Albrecht, have noted an important phenomenon in today's complex business and service organizations: The frontline staff who directly serve the customers or users are the most important to organizational success.

Management must itself be a service. In other words, management needs to assist, train, and support those who serve the customers and users. Management should provide needed resources to those who serve the customers or clients.

This is perhaps best shown by the "pyramid of service," which is the reverse of most organization charts. From top to bottom, the levels of importance are these:

1. **Customers**, clients, users, beneficiaries
2. **Frontline** service providers (staff)
3. **Supporting** managers (resources and coaches)
4. **Executives** and senior management (system designers)

Other Thoughts

Tom DeMarco and Timothy Lister coined the term "teamicide" in their excellent book *Peopleware: Productive Projects and Teams* (New York: Dorset House Publishing, 1987). In their book, the authors identify another set of problems that may be caused (or tolerated) by managers.

These "teamicide techniques" include few surprises but are worth a reminder to us here. In their view (and mine), these rather commonly tolerated problems can be very destructive to teams:

Ø Defensive management

Ø (Unnecessary) bureaucracy

Ø Physical separation (of team members)

Ø Fragmenting people's time (on many different projects)

Ø Quality reduction of product (or project)

Ø Phony (or unrealistic) deadlines

CHAPTER 10

DIFFERING STYLES, PERSONALITIES AND CULTURES

Δ Δ Δ

It has been my privilege to assist many management groups that are highly multicultural and diverse in their membership. These groups have included minorities, women, and people of all ages with varied experiences and educations. More recently, these groups have been of mixed nationalities, including European and Asian as well as American managers. As challenging as these groups are for their consultant, they are even more challenging for one another!

Critical skills for all kinds of teams in the years ahead will be understanding, appreciating and working well with colleagues, customers, and partners who are quite different from ourselves. These differences will be many, and they include the following:

Δ Country of origin	Δ Culture and traditions
Δ Educational background	Δ Working experiences
Δ Gender and orientation	Λ Personality types

Diversity and Strength

When group members appreciate their differences, they have several strengths and advantages that may materialize in their work together. When other factors such as education and experience are equal, these groups are generally more lively, creative, and wide-ranging in their thinking than are less diverse groups.

> A word about **culture**: Culture is the sum of "how we do things around here." It is influenced by traditions, myths, history, and heritage. Culture is manifested in norms of behavior (unwritten rules) as well as in rewards and sanctions.

Coming from a wider background of cultures, subcultures, and traditions, members of diverse groups have more divergent perspectives. They are less likely to take things for granted and will probably have a broader knowledge base.

This strength of diversity, however, may be and often is also a weakness. A broader knowledge base, clash of norms, not taking things for granted, and challenging one another's thinking can make members uncomfortable with one another.

A very basic need for effective teams is a tolerance for differences in people and their perspectives. But more than tolerance is required for high-performing teams. Members must come to appreciate their differences for the value they bring to the team despite any discomfort that may come with it.

Personality Differences

Beyond differences in culture, norms, and perspectives among members, there are some other important differences among people that must be recognized and appreciated in order to be effective in one's work and life.

> Recognition of personality type differences goes back many centuries. In 400 BC, Hippocrates spoke of the four temperaments, which he identified as follows:
>
> Δ *Sanguine:* The extrovert, optimist, talker
>
> Δ *Melancholy:* The introvert, pessimist, thinker
>
> Δ *Choleric:* The extrovert, optimist, doer
>
> Δ *Phlegmatic:* The introvert, pessimist, watcher

Most people have aspects of several personality or temperament types, adding both spice and complexity to the task of working together.

Group members owe it to themselves and their colleagues to develop a good understanding of personality or temperament types. Many excellent resources have become available in recent years. Here are some of my favorites:

Δ **"Understanding Psychological Type to Improve Project Team Performance,"** a paper by Gordon Culp and Anne Smith, as published in the *Journal of Management in Engineering* by ASCE, Reston, VA, 2000; www.asce.org.

Δ *What Type Am I?: Discover Who You Really Are*, by Renee Baron, published by Penguin Books, New York, 1998.

Δ *Type Talk: The 16 Personality Types that Determine How We Live, Love and Work*, by Otto Kroeger and Janet M. Thuesen, Dell Publishing, a division of Random House, 1989; www.randomhouse.com.

Δ *People Types & Tiger Stripes: A Practical Guide to Learning Styles*, 3rd ed., by Gordon Lawrence, published by the Center for Applications of Psychological Type, Inc., Gainesville, FL, 1997; www.capt.org.

Δ *Personality Plus: How to Understand Others by Understanding Yourself*, revised and expanded ed., by Florence Littauer, published by Fleming H. Revell Co., a division of Baker Book House, Grand Rapids, MI, 1992; www.bakerbooks.com.

Δ *Gifts Differing: Understanding Personality Type*, 2nd ed., by Isabel Briggs Myers with Peter B. Myers, Davies-Black Publishing, a division of Consulting Psychologists Press, Inc., Palo Alto, CA, 1995; www.cpp-db.com.

All but one of these books are based on the basic personality types identified by Carl Jung and serve as excellent resources. Littauer's book, *Personality Plus*, is more basic but provides excellent background material for any other reference or method of understanding personality differences.

Other Differences

Many other systems, profiles, and typologies exist for noting differences in various important aspects of working together, such as:

Δ Approaches to conflict

Δ Social styles

Δ Methods of problem solving

Δ Internal motivation

Δ Ways of processing information

Δ Leadership styles

Any of these is quite useful in appreciating the natural differences in people, some of which are important because they frequently pop up and/or cause irritation or annoyance among colleagues, partners, or customers. Some of the most frequently observed, heard, or believed differences among people working together are shown in *Figure 10.1.*

51

Internal, own goals and standards	----**Locus of Control**------	*External, others' goals and standards*
Preference for being with other people	--------**Sociability**---------	*Preference for being alone or with one person*
Likes details and sensory data	--------**Perception**----------	*Likes imagination and concepts*
Tends to trust logic and standards	----------**Decisions**-----------	*Tends to trust values and feelings*
Decisions, plans, and schedules	----**Copes by Way of**------	*More information, options, flexibility*
Style, flair, appearance	-------**What Counts**--------	*Substance, content, value*
Conflict is OK, let's argue	-------**Disagreement**-------	*Be harmonious, don't fight*
Reactive, responsive to others	-----**Source of Action**-----	*Proactive initiatives and actions*
Work, work, work	----**Preferred Activity**----	*Take time to enjoy*
Do it quickly	-------**About Work**--------	*Do it well*
Be calm and cool	----**Showing Emotion**-----	*Let it all hang out*
Pessimistic, cautious	--------**World-view**---------	*Optimistic, enthusiastic*
People	--**Focus of Management**--	*Tasks*
Directing	--------**Manages by**--------	*Leading*

Figure 10.1. Ranges of Differences in People
Noted Frequently in Organizations and Groups

(These are in no particular order, and there is no intended correlation between the left and right side of one aspect and the left and right side of another.)

Some personality types are normally distributed in the general population, while others are skewed to one side or the other. In some organizations, types tend to cluster because of the selection process and/or training and enculturation process in the group or organization.

Racial, Cultural, and Gender Differences

While significant progress has been made in understanding and bridging racial, cultural, and gender differences in North America (to varying degrees), much work remains to be done in almost all organizations.

Stereotyping is a natural human tendency, even helpful sometimes in that lumping groups of things, events, or situations together can simplify certain tasks. It would be terribly wasteful if we had to learn and relearn that rocks are hard, fire is hot, and so on.

***

Problems that arise from stereotyping include the following: (1) We often apply our skills at categorizing and lumping things together to people who are only superficially similar. (2) We accept what others have said about groups without using facts readily at our disposal. (3) When we do look at what is true about individuals, we consider them to be "exceptions to the rule."

***

These last three items constitute the practice of prejudice, or prejudging (based on superficial or inaccurate data, assumptions, or hearsay). Rob Terry, author of *For Whites Only* (1994), notes that the power to act or influence others (to exclude people, to hire or not hire, to pay well or poorly, to promote or not) coupled with prejudice equals racism, sexism, and other types of discrimination, like so—

(POWER X PREJUDICE) = (RACISM, SEXISM, AGEISM, ETC.)

More on Stereotyping and Prejudice

Most of us are **not** aware of the hundreds of subtle myths, half-truths, beliefs, and nontruths we have inherited (unexamined) from the important people in our lives, including parents, grandparents, siblings, play- and schoolmates, and the media. Even our churches and schools have contributed (perhaps unwittingly) to these learned prejudices.

Gordon Allport's studies of prejudice development (*The Nature of Prejudice*, 1979) showed that 70% of the people studied attributed learning about prejudice primarily to parents or family. He found that such prejudice in children was learned considerably later in life than when children learn most other important life attitudes (from 0 to 6 years of age). He also observed that, once learned, we tend to keep our prejudices and stereotypes intact in the following ways:

Δ **Lack of data:** Whether through lack of experience with certain groups of people, our own internal filters, or both, we simply do not "see" that individuals are not like our stereotypes of them. The evidence goes unobserved or undigested

Δ **Exceptionalizing:** We believe that specific individuals we know are "exceptions" to the "rule" (which we continue to believe is true).

Δ **Re-fencing:** When finally confronted with massive data about either individuals or a group that is contrary to our prejudice, we may fall back on some other aspect of our prejudices.

It is in this area that we are most likely to misuse our intellectual abilities to kid ourselves. We must be willing to "hear" feedback, both verbal and nonverbal, in order to gain the needed knowledge despite our biases.

Dealing Effectively with Differences

It's easy to be prescriptive about any differences in people, to believe that one aspect is right and another wrong, but this is not necessarily so. It seems to take all types to make the world work, no matter how uncomfortable these differences may feel to us.

Psychology, behavioral science, and management studies have given us some clues about how best to deal with these and other differences in people:

Δ **Recognize** that diversity in a group not only adds to the conflict potential but also adds to the group's potential for creativity and better solutions as well as to our own potential for learning and development.

Δ **Consider** the possibility that when another individual is making you uncomfortable (although not in any way threatening to you), it may be due to their challenging some belief, opinion, bias, or prejudice of yours. This may provide you with a new perspective or new information or perhaps remind you of something you've disliked in yourself.

Δ **"Hang in there"** in situations in which differences are difficult for you, particularly if you have a sense of progress. Consider asking for outside counseling or facilitation. Be open to feedback from the other parties.

Δ **Put yourself in their shoes**. Try to experience their side of the situation. How does it sound, look, feel now from **their** perspective?

Δ **Mentally move away** and try to become a neutral observer or third party (become your own consultant). How does it sound, look, and feel from **that** perspective?

Δ **Change their behavior** toward you by changing some of **your** behavior toward them. No one can directly change another person. However, individuals will certainly change in response to a change you make, if the change is positive and visible (to them) and it is held long enough to be noticed and "tested" for genuineness.

> **For example:** Sally, vice president of marketing, was often confused and upset whenever CEO Ralph reviewed her plans or work critically. Sally made two changes. First, she now asks Bob (vice president of human resources) to critique her work before her meetings with Ralph. Second, she now notes on a pad each of Ralph's points as he makes them.

> As a result, Sally is less apprehensive (before meetings) or upset (during meetings) and Ralph is more respectful in his criticism (in response to Sally's changes).

Helpful Resources

Δ *The Nature of Prejudice: A Comprehensive and Penetrating Study of the Organization and Nature of Prejudice*, unabridged 25th anniversary ed., by Gordon W. Allport, Perseus Publishing, Cambridge, MA, 1979; www.perseuspublishing.com.

Δ *The Content of Our Character: A New Vision of Race in Amercia*, by Shelby Steele, Harper Trade Publication, New York, 1991.

Δ *For Whites Only*, by Robert W. Terry, William B. Eerdmans Publishing Company, Grand Rapids, MI, 1994.

In Terms of "People Differences" …

Δ   What types of situations are uncomfortable for me?

Δ   What groups or individuals tend to be uncomfortable for me?

Δ   What personalities or types of people make me uncomfortable?

Δ   How do I react or behave in this/these situations?

Δ   Are there any changes I might consider making in my …

   Δ   Skills?

   Δ   Knowledge?

   Δ   Attitudes?

CHAPTER 11

RESPONSIBILITIES OF INDIVIDUAL TEAM MEMBERS

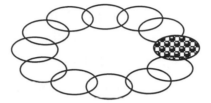

Price Pritchett has written a useful, practical, and compact handbook for team members that addresses a great need in the teamwork literature. As he says in *Teamwork: The Team Member Handbook* (Dallas, TX: Pritchett Publishing Co., 1997), he believes that "the secret to teamwork lies in the team members."

That's true, he says, whether we're talking about basketball, surgery, fire fighting, music, or drug busts. Here's a list of suggestions for people who want to be seen as **effective team members**, based on his work and mine.

- **Know your job**, and do your best at it.
- **Continue to develop** your work skills.
- Support and **use the diverse talents of the team**.
- Work hard at **effective communication**.
- **Keep the big picture in mind**, so you can …
- **Help others** who need an assist from a teammate.
- **Bring real problems** to the team's attention.
- Build up your teammates; **give recognition**.
- **Watch your ego**; there's no "I" in teamwork.
- **Be a good sport**; use humor in positive ways.
- In public, **support team and leader decisions**.

Behaviors to Avoid

Some individual behaviors are a problem for the team and the leader and need to be avoided by each member. When they occur, and they will, they need to be brought to the offending member's attention.

There are probably thousands of **team-busting behaviors**, but here's a short list of frequent abuses to watch out for:

- Being **stubborn** to a point of frustration
- Being **sarcastic**, cynical, and generally negative
- Covertly **badmouthing** others, creating rumors, etc.
- Engaging in **horseplay**, nonchalance, wisecracks
- Seeking personal **recognition**, self-promoting
- Being **indifferent**, aloof, nonparticipative
- Riding your **personal agenda** into the ground
- **Manipulating** the group in various ways

Stretching Your Skills (Leadership)

> More than a job title, role, or position, **leadership,** it seems to me, **is a set of skills and facilitative activities**.

As such, all team members must contribute leadership to their group at various and appropriate times. Following are some leadership skills and abilities every member should work toward developing:

- Take a broader view of your organization, including the needs and priorities of other parts of it.
- Value openness and honesty and learn to use them with tact and gracefulness.
- Polish your personal skills of listening, negotiating, managing conflict, and communicating clearly.
- Learn the group skills of planning, team development (see Chapter 7), and problem solving (see Chapter 8).
- Practice the skills of delegation, giving recognition and follow-up where warranted.
- Don't pretend to know what you don't know, and don't pretend to understand what you don't understand; ask questions and learn.
- Remember, many organizations are constrained in their growth by a lack of enough excellent leaders or potential leaders.

CHAPTER 12

TEAM DEVELOPMENT THROUGH FEEDBACK

Groups and teams need to be effective in whatever they're doing; they need to achieve their mission or purpose. Just as important, every team or group needs to be seen as effective by its

- ↻ **clients** or customers
- ↻ **users** and beneficiaries
- ↻ **owners** or stakeholders
- ↻ **regulators** (if appropriate)
- ↻ **group**/team members
- ↻ **colleagues** in other groups

How a group is viewed or regarded by others is essential information if the group is to continue to be supported, provided with resources, and have good future work opportunities.

This is true even in the public sector (e.g., schools, agencies, hospitals, utilities)—eventually. It may take longer, but sooner or later public sector groups will be held accountable by those they supposedly serve.

What is Useful Feedback?

Private sector groups, for the most part, know a good deal about how well they are doing by means of almost "automatic feedback," such as

- ↻ changes in sales, demand, backlog

- ↻ changes in complaints, returns, fixes

- ↻ changes in position relative to competitors

- ↻ changes in enrollments, transfers, etc.

Although this is useful and important information, often something more would be helpful in terms of

- ↻ **why** things are as they are

- ↻ **what**, specifically, would be more valued or valuable

- ↻ **new features, services, or products** that customers need

- ↻ **how** customers might feel about something they "don't know they need"

- ↻ **comparisons** of your performance versus that of other providers

Market research findings often indicate that many private sector organizations get their best ideas for new features, services, and products from their clients and customers.

My experience strongly suggests that product and service providers could learn a great deal more from their customers than they take advantage of. So—a competitor beats them to it.

Getting Useful Feedback

Believe it or not, getting useful feedback is the easy part! **Why?** Because the group or team needs to develop and design the feedback process, then obtain and analyze the resulting feedback!

How feedback is obtained is incredibly important because—

- ↻ If the feedback is to be worthwhile, it must be used to improve!

- ↻ If the feedback is to be used by the group or team, members must be intimately involved!

This way the group develops "ownership" in the resulting feedback.

There are drawers full of great survey data, collected by very skilled and professional survey firms or individuals for their clients. Most of it becomes nothing but file-fillers. Such surveys and feedback seem to elicit little besides rationalizing and excuses because **the people who receive it don't "own" it**.

Feedback Processes

There are many ways for groups and teams to elicit and gather useful, **helpful feedback**. Here are some that are often used (and need to be fine-tuned by the group or team):

- ↻ **Focus groups** of people from one client organization

- ↻ Focus groups of people from several client organizations

- ↻ **Reverse seminars**, given by clients, for you and your team

- ↻ **Telephone surveys**, which get more and better data than written surveys (snail mail or email)

- ↻ **Surveys** asking qualitative questions

- ↻ Surveys asking evaluative questions

- ↻ Surveys asking clients what they think is important

Some Good Questions

Over the years, many of our clients have asked their clients excellent questions to gather feedback on performance as well as ideas that may be useful for their future. Here are some of my favorites:

- ↻ How do you compare us with other service providers?

- ↻ In a service such as we provide, what three things are most important to you?

- ↻ On a scale of 0 to 5, how well do we do on those three things?

- ↻ What do you consider our strongest points of service?

- ↻ What do you dislike about our current services?

- ↻ What would you like to get from us that we either don't do or don't do well?

- ↻ What do you think are key trends in our market?

- ↻ What do you or will you want from us in the future?

CHAPTER 13

GETTING YOUR GROUP OR TEAM UNSTUCK

In the natural course of events, difficult circumstances arise. Surprises and changes, sometimes nasty, seem to pop up, usually at the worst possible times. "Murphy" was definitely an optimist. **Teams do get stuck**.

Critical Events

I have identified a number of critical events or situations in which groups or teams can be adversely affected. Some of the most common or frequent critical events for teams include the following:

Losing a leader

Gaining a leader

Losing a member(s)

Gaining a member(s)

Problematic leader or member

Changing charter or mission

Changing technology

Shortage of resources

Increasing competition

Change of ownership

Merger or acquisition

Loss of market or users

Events such as these (and others) can result in a group or team getting "stuck." In these situations, the group may feel unequal to the task or may be unsure of itself, the task, how to go about it, or all three.

What the Leader Can Do

In difficult times like these, many leaders and managers instinctively react with authority and strong direction despite their inner lack of clarity and confidence. Often, that's not the best reaction for the group or situation. Instead, this may be the time to rely on the (almost) surefire dilemma-solving strategies for leaders and managers:

 Ask the group/team for help.

 Facilitate their discussion.

One might initiate such a discussion by saying, "I know we have a problem here with _____, but I'm not clear about what to do. Let's discuss it and see what the group/team's collective wisdom is." This is the time to use all the resources of the team.

At such a time, one person (not necessarily the leader or manager) needs to ignore the problem itself and instead **facilitate the process** for the group. Whoever the facilitator is, he or she needs to be very disciplined and tend to the process only, trusting the group or team to handle the data.

It helps to have a simple but effective group problem-solving model to follow during such times. If there is no model in place, people tend to spend too much time and energy on arguments, positions, and pet solutions and too little time on understanding the situation. Consider the team tools discussed in Chapter 8.

When the "Problem" is a Member

Sometimes a team or group can be badly "stuck" when a member has a serious problem that impacts both his or her performance and the team's performance.

Prevention is the best remedy. Select team members with care whenever selection is an option. Assess all aspects that impact performance:

Knowledge—Easy to assess, **easy** to change

Attitude, skills, habits—Difficult to assess, **difficult** to change

---

If the problem is one of **skills or knowledge**, the leader and team **may** be able to help through on-the-job coaching, special training, or motivation for learning.

If the problem is one of **attitude or habits**, the leader and team **may** be able to motivate the member by means of helpful personal feedback, using the following specific "recipe":

- What performance behaviors you observe
- How that impacts the team or others
- What you prefer instead of what you observe

Example:

- "You've been 20 to 30 minutes late for operating meetings this month."

- "Eight other people are delayed in their work if we wait" or "We lack your input if we don't wait."

- "We would like you to be the first person here for the next five meetings."

Sometimes these efforts to help may not achieve the needed improvements. It then becomes the leader's responsibility to change the status quo through the usual appropriate personnel procedures for the particular organization, such as

- counseling off-line
- disciplinary action(s)
- transfer or termination

Consider Bion's Findings

Wilfred Bion was a British soldier in World War I and a physician in World War II who studied groups. He came to some controversial yet useful conclusions that are little known but very helpful for business organizations.

Bion noted that groups are either "on" or "off"; that is, they are either **working** or **not working**. Some of the important differences are shown in *Table 13.1.*

Table 13.1. What's Happening When Groups Are "ON" or "OFF"

	On ... Working
Issues:	✓ Effective groups deal well with their issues and face them despite member anxiety. ✓ Effective groups have many issues but keep whittling them down.
Communication:	✓ Members of effective groups work very hard to be **both** - sensitive/tactful - candid/honest ✓ People speak for themselves, using "I" messages (e.g., "I feel ...," "I believe ..."). ✓ People are slow to "understand." They ask and listen to the answers in order to really understand!

Ground Rules:	✓ Effective groups have a common purpose, a shared fate, and explicit ground rules such as
	
	  - "We learn to face and cope with problems, issues and discontent."
	
	  - "No surprises; let everyone know you plan to raise an issue."
	
	  - "We revisit our expectations." Most problems are due to un-met expectations, or unrealistic expectations.

Leadership	✓ The leader's role is facilitative, supporting the ground rules, focusing the group on its issues and working on them, ensuring that each member is valued and heard.

✕ **In**effective groups duck their issues, refusing to face them as a group yet discussing them covertly in a gossiping manner.

✕ **In**effective groups have truckloads of issues that keep piling up!

✕ Members of **in**effective groups are **either**

    - sensitive/tactful   - candid/honest

Either of which alone has the tendency to shut down real discussion and open communication.

✕ People use **in**direct forms of communication:

- Speaking for others
- Invoking "expert" opinions
- Using questions for statements ("don't you think ...")

✕ **In**effective groups have norms that are set by the behaviors of the dominant members.

✕ These norms are usually discovered whenever someone accidentally breaks one.

✕ The norms are generally supportive of comfort and predictability rather than group effectiveness.

✕ The leader is expected to take on the issues privately so the group doesn't have to. This routine is more comfortable for the leader for a while, but in the long run it hurts both the leader and the group.

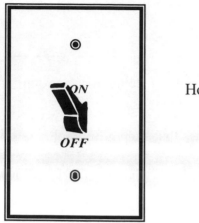

How often is my group …

    ✓ "ON?"

    ✗ "OFF?"

What evidence is there for that?

    ✓ "ON" …

    ✗ "OFF" …

CHAPTER 14

CONFLICT: HANDLING DIFFERENCES
IN GROUPS AND TEAMS

The Avoidance of Controversy

Controversies are a natural and desirable part of any decision-making or problem-solving situation. When managed constructively, they are extremely valuable.

Controversies are absolutely necessary *if an organization is to make effective decisions and competently solve problems to maintain its viability and effectiveness.*

Yet too few organizational members accept conflicts among ideas, information, and conclusions as being desirable, and almost none attempt to stimulate it. There are at least three reasons why:

➤➤ *First, controversy is not a standard procedure in most decision-making and problem-solving situations and, therefore, too few organizational personnel understand it. There is insufficient knowledge and understanding of the procedures involved in controversy and the advantages and potential constructive outcomes that can result from disagreement.*

➤➤ *Second, most organizational personnel seem to lack the interpersonal skills and competencies needed to stimulate controversy and ensure that it is managed constructively.*

➤➤ *Third, there seems to be considerable fear and anxiety on the part of most people in conflict situations. Disagreement is often viewed as stressful and threatening.*

—David W., Frank, and Roger Johnson[6]

---

[6] David W. Johnson, Frank Johnson, and Roger Johnson, "Unlocking the Potential of Group Decision Making through Constructive Controversy," *Creative Change*, fall 1982, published by the Association for Creative Change.

The Problem of Conflict

Many group leaders and members say they have a terrible problem with conflict in their groups, meetings, and discussions.

Conflict is indeed a significant problem in many types of organizations, including churches, government services, hospitals, and healthcare teams. Conflict management professionals (e.g., mediators) as well as manufacturing, engineering, design, construction, and mining organizations also experience a great deal of conflict.

Conflict is particularly wasteful and dysfunctional when it is unmanaged or poorly managed or when it produces **heat** (hassles and disruptions) rather than **light** (positive outcomes). In my work with organizations of many kinds, I have seen significant costs incurred due to poorly managed conflicts. These costs are real and often measurable:

➤ **Poor-quality decisions are made.** When opinions and decisions are influenced by personal comfort, ego, self-interest, or turf protection, then the organization, its clients, users, customers, and beneficiaries suffer.

➤ **Staff motivation suffers.** Morale is impacted, which then causes a loss of focus, commitment, and productivity; gossip, wasted time, and worry increase.

➤ **Employee turnover increases.** Skilled human resources are lost and replacements cost more in both lost time (never recovered) and dollars.

➤ **Working hours are lost.** In an effort to avoid an unpleasant environment, people stay home more and are impacted by stress-related illnesses.

➤ **Healthcare costs increase.** Costs associated with illness, stress, and lowered morale and resistance increase. Drug and alcohol abuse are more likely as well.

The Real Problem

There is no doubt a real problem with conflict. However, the problem isn't that conflict is so prevalent. As people with diverse family backgrounds, educations, experiences, ethnic origins, values, temperaments, and responsibilities try to work together, it's almost a miracle that two people ever agree!

> Research and common sense suggest that disagreement, controversy, and conflict are inevitable. Because we can't remove conflict, outcomes will depend on the ways in which we deal with it (or not), handle it (or not), or manage it (or not).

There are hundreds of ways to deal with conflict that are not functional, even though they may seem to "work" for the moment. Following are some of the most frequently used approaches:

- Blaming or making others "wrong"
- Fearfulness—avoiding issues, ducking
- Sarcasm and ridicule
- Indirectness—using games and hidden agendas
- Coercion—using power plays and intimidation
- Filibustering—talking people to death
- Displaying anger and angry behaviors
- Sabotage, going underground

These apparent remedies have something in common: They are all forms of avoidance in a variety of disguises *(Table 14.1)*.

Table 14.1. Forms of Avoidance

"SOFT" FORMS OF AVOIDANCE	"TOUGH" FORMS OF AVOIDANCE
- Fearfulness - Indirectness - Filibustering - Sabotage	- Blaming others - Sarcasm, ridicule - Coercion, power plays - Angry behaviors
Goals 1. To avoid direct mutual discussion of issue(s) and resolution 2. To avoid any personal discomfort or difficulty for self/other(s)	*Goals* 1. To avoid direct mutual discussion of issue(s) and resolution 2. To win over the other(s) as quickly as possible

The major problem with conflict and disagreement then is not so much that it exists, rather that our predominant ways of dealing with it are avoidance of one kind or another.

Using Conflict Productively

Research shows over and over again, from Blake and Mouton (*The Managerial Grid*, Gulf Press, 1964) to Dean Tjosvold (*The Conflict Positive Organization*, Addison-Wesley Longman, Reading, MA, 1991), the need for a productive approach to conflicts. These books are excellent references for understanding and using conflict.

The most useful single and compact reference for managers on this vital topic is an excellent paper by Amarjit Singh and Demetres A. Vlatas (members of ASCE) called **"Using Conflict Management for Better Decision Making,"** *Journal of Management in Engineering* (Vol. 7, No. 1, paper 25434, pp. 70–82, January 1991; Reston, VA: ASCE Press).

In it, they identify five approaches for handling conflict: forcing, compromising, withdrawing, smoothing, and problem solving. Only problem solving produces positive results a very high percentage of the time; this is followed by compromise seeking *(Figure 14.1)*.

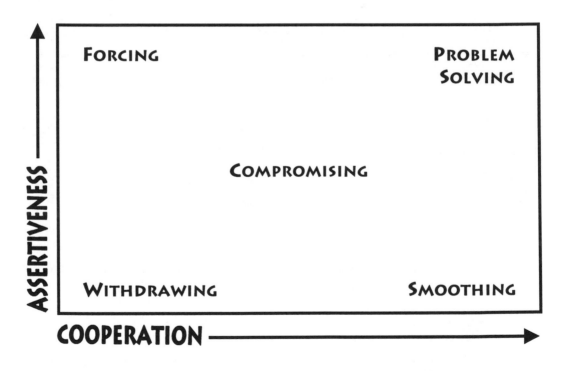

Figure 14.1. The Five Approaches for Handling Conflict

(These approaches are characterized by increasing assertiveness and/or cooperation.)

The key to using conflict and producing quality decisions and creative solutions from our natural conflicts is what we call the Problem-Solving Approach. This approach is typified by the following:

➨ Being **patient** with the process

➨ Being **persistent** toward resolution

➨ Showing **respect** for others (even if it isn't initially returned)

➨ Searching, jointly or alone, for **alternatives** that may satisfy all sides

➨ Working for **creative options** that are not yet on the table

➨ Managing one's **own actions** (despite one's feelings)

Referring to stage IV of the four stages of group development (*Table 6.1*), high-performing teams have a great deal of conflict that often looks like problem solving. Conflict doesn't go away and it isn't avoided; instead it is used and well managed.

It Only Takes One

It only takes **one**. One group member who takes the problem-solving approach can bring resolution to a roomful of sturdy battlers. It's not just theory—it works!

Although it only takes one, it does take some doing. Sometimes it may seem like trying to nail jelly to the wall. But if one person can manage his or her emotions and focus on the process, conflicts yield to creative solutions.

In recent years, more and more of the lay psychology books, tapes, and workshops have promoted various avenues or approaches for mental and emotional well-being. Common to the best of them is the theme of self-management; which is the first and perhaps most difficult aspect of management. Prominent leadership author Warren Bennis notes that management of self is critical; without it, one can do more harm than good.

The emotions most often in need of management are fear and anger, or their milder versions, anxiety and irritation. Trying to ignore such feelings almost always results in failure; we fool only ourselves—if indeed we fool anyone. The keys to managing these and other strong emotions are these:

➨ Pay attention, notice your emotion(s).

73

➡◄ Acknowledge your emotions, and accept them.

➡◄ Behave based on your intentions, not your emotions.

(In some cases, certain strong emotions such as caring or confusion are appropriate to share because they lead to trust, clarity, and progress.)

Summary

Conflict in groups is natural, inevitable, ultimately unavoidable, usually manageable, and capable of producing creative solutions if well managed.

The problem-solving approach requires patience, persistence, respect, searching, creativity, and managing oneself despite strong emotions.

Management of oneself includes noticing emotions, acknowledging and accepting emotions, and behaving based on intentions not emotions.

CHAPTER 15

OVERCOMING SEPARATION IN TIME OR SPACE

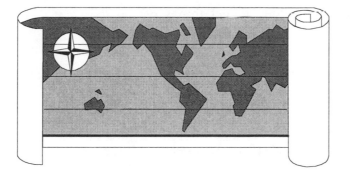

Groups such as project teams, practice networks, and marketing groups are becoming ever more dispersed and distributed.

For example, on a current project for a major water utility, I'm one of 10 consultants assisting the primary consultant. In addition, the primary consultant's technical manager, project manager, and key technical people are in five different locations.

Electronic Helpers

Staying connected is made easier by communication and information technology, including

* cell phones and beepers
* email (if used appropriately)
* faxing (especially graphic information)
* intranets and company and project web sites
* teleconferencing/videoconferencing
* computer graphics—two- and three-dimensional
* direct links between service providers and clients

More is Needed

> Many organizations have discovered that groups and networks need some "face time" (time **together**, face to face) to create a start-up team! Electronic media are very efficient at information transfer, but they are not as effective for true communication!

In my earlier example of the major utility, the project group (about 20 in all) had a kick-off meeting to get things off on the right foot. Included were several people:

✦ Water Utility: Project manager and key technical people

✦ Primary Consultant: Project manager, technical manager, and key members from other offices

✦ Subconsultants (10): Key person from each

With reference to the stages of group development (see Chapter 6), the **objectives of the kick-off meeting** were to accomplish the following:

✦ **Move** from stage I to **at least** early stage III

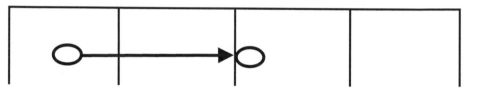

✦ **Agree** on project mission, major goals, and milestones

✦ **Create** needed procedures, including:

   ✦ Communication protocols (who, when, where, etc.) for one-on-one and regular meetings

   ✦ Communication ground rules (how, etc.), similar to team ground rules (see Chapter 8)

   ✦ Conflict resolution process for handling the inevitable issues and disagreements

✦ **Become** more familiar with and comfortable with one another's personalities and unique roles in the project

Another Example

Several years ago, a manufacturing client wanted to develop a plan and build a team of it's in-country divisions and partners in Asia. Again, the first step was to get all the key people together face to face.

We chose a neutral site with a reasonably central location to minimize travel time and costs. These key executives and co-owners were very appreciative. The resulting plan and teamwork provided a strong base for future easy communication by fax, email, phone, and intranet.

Even though English was not the primary language spoken by many team members, all agreed to use English as their common communication vehicle. As one member said, "English is easy to learn well enough to be understood—even when spoken badly."

Bottom Lines

✤ **Face-to-face** work is important to any group or network early on.

✤ **Team tools** (see Chapter 8) are even more critical when the group is dispersed in time and/or space.

✤ **The stages of group development chart** is another helpful tool (see Chapter 6) to enable teams to

 ✤ get started more easily

 ✤ establish a benchmark

 ✤ monitor their progress

 ✤ talk about teamwork

CHAPTER 16

SHARED LEADERSHIP IN GROUPS AND TEAMS

> *Over the years I have joked with my colleagues that perhaps what is really needed is a course on "**dynamic followership!**" After all, what does a leader need to know other than how to create, nurture, support, and encourage followers? There are 10,000 courses available on leadership, but no one even mentions followership.*
>
> *I think what leaders need is the ability to do for others (to equip, support, guide, empower, and embody). What do leaders need to know? It seems to me that leadership has something to do with the effective interaction among leaders and followers, and somewhere near the center of effectiveness is the concept of servant.*
>
> —John C. Bryan[7]

The qualities needed for effective leadership have been argued for centuries. Many recent researchers and authors have made significant contributions worth having close at hand on your leadership and management bookshelf. Four of my personal favorites include the following:

- *First, Break All The Rules: What the World's Greatest Managers Do Differently*, by Buckingham and Coffman, published by Simon & Schuster, New York, 1999.

- *How to Solve the Mismanagement Crisis: Diagnosis and Treatment of Management Problems*, by Ichak Adizes, published by The Adizes Institute, San Diego, 1979.

- *On Becoming a Leader*, by Warren Bennis, Perseus Publishing, 1994; www.perseuspublishing.com.

- *Servant Leadership: A Journey Into the Nature of Legitimate Power and Greatness*, by Robert K. Greenleaf, published by Paulist Press, 1983.

---

[7] John C. Bryan (Bryan Weir Bryan Consultants, Toronto, Canada), **"Dynamic Followership,"** *Journal of Creative Change* (formerly *Journal of Religion and the Applied Behavioral Sciences*), winter 1987.

Implications for Leadership

Adizes' research contribution is unique in its significance for leaders and managers everywhere:

 No single individual has all the required skills or capabilities for leadership and management!

 A leader is fortunate to have half the qualities needed to manage his or her organization!

Bennis' research contribution is also unique and useful. He found that **successful** leaders possess the following qualities:

 Although able and dedicated, they have real limitations.

 They are very aware of their own strengths and limitations.

 They draw on the needed strengths of others to fill in the voids.

The implications of the work of both Adizes and Bennis are profound. Coming at the matter of leadership from two very different, yet practical, research perspectives, here's what we believe their work points to:

For any group, team, or organization to have all its needs for leadership and management fully met, **leadership must come from more than one person**; it must be shared.

Leadership/Management Functions

The various capabilities of leadership/management can be boiled down to the irreducible minimum. Actually, the capabilities may be as few as two and revolve around to the needs for both stability and change and the balance between them:

 Management: Concern for the ongoing service and production by people in the group or team and the systems they use

 Leadership: Concern for the necessary change and development to improve their services, quality, and effectiveness

A somewhat longer list of the essential functions of leadership and management might look like this:

1. **Core competencies:** Intimate knowledge and skills in the core competencies of the unit's or organization's products or services and the ability and willingness to share and teach others

2. **Organization:** Abilities that help create order from chaos and develop the systems required to handle large or complex projects and/or organizations, including masses of detail

3. **Entrepreneurship:** Practical creativity that produces new or improved products and services and/or enthusiasm to capture the interests of others, including colleagues, clients or customers

4. **Human relations:** Understanding others' feelings, values, needs, and interests and the ability to influence people with differing interests to work together toward common goals

5. **Growth and development:** The drive and skills to develop people and organizations to be all they can be; help them to grow and improve as necessary to survive and prosper

6. **Financial/business:** Whether public or private, profit or nonprofit, every organization needs skills in pricing, budgeting, negotiating, contracting, funding, tracking, and other aspects of cost-effective management

This list of essential functions of leadership and management has some overlap, and it would look different if sliced from any other perspective. Even this short list demonstrates that **only rare and remarkable individuals are strong in all six functions of leadership and management.**

So perhaps the most important thing to know about leadership is that it must be shared for the sake of organization effectiveness. In other words, the need for teamwork begins at the top of any group, unit, or organization.

An Application

When assisting clients in their strategic planning and/or organization design efforts, I sometimes ask them to do a "management functions inventory." When completed, it often looks much like *Table 16.1*.

In this example, all the functions are covered by several people. Core competencies are the most heavily represented. Note that growth and development are the least represented for this somewhat typical group.

Table 16.1. Sample Management Functions Inventory

CORE COMPETENCIES	ORGANIZATION ABILITY	ENTRE-PRENEURSHIP	HUMAN RELATIONS	GROWTH & DEVELOPMENT	FINANCIAL/ BUSINESS
Bill			Bill		
Ralph	Terry	Terry	Ralph	Terry	Jerry
Susan	Ralph				
Wade	Jerry	Susan		Susan	Wade
Tim		Wade	Tim		
	Karen				Karen
Very strong here	**Pretty strong here**	**Fairly strong here**	**Fairly strong here**	**Not so strong here**	**Fairly strong here**

This exercise usually accomplishes three important things for the group:

 Members are encouraged to examine and appreciate various capabilities in others.

 Members can more clearly see how dependent members are on one another as well as the interdependence of the group.

 An empty or nearly-empty column indicates a need that should be addressed.

Self-Led Teams

Several progressive companies have made good use of leaderless teams, sometimes known as **self-managing work groups**. This leadership innovation was first tried by Procter & Gamble in several new plants.

As mentioned earlier, leadership must exist not necessarily as a traditional manager or executive but as a set of skills and behaviors, probably shared, to enable and support the group or team in its work.

Within such autonomous work groups, each of the six essential functions must be handled by someone as the needs arise. They are carried out by those in the group with the skills to do them and with the support of the group.

Such a group would need to be at a high stage of team development (stage III or IV) in order to function well in all areas. The abilities of members would be appreciated and used by the group, particularly if group rewards are for results and performance.

Teams like these provide excellent opportunities for full development of members; they enable people to contribute all of which they are capable. Such teams also provide many good models of leadership and management as opposed to one model or one right way.

Leadership Summary

Leadership/management is

 A set of skills and abilities

 Needed in every group, team, or organization

 Concerned with both stability and change

A set of six essential functions and abilities:

   Core competencies

   Organization (and systems)

   Entrepreneurship (creativity and enthusiasm)

   Human relations (common goals)

   Developing people and organizations

   Financial and business management

 Seldom provided fully by just one individual

 Dependent on teamwork in order to be effectively shared among group members

CHAPTER 17

TEAMS ARE THE BEST DEVELOPMENT VEHICLES

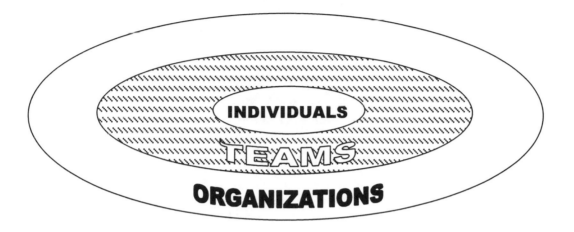

Any natural team or work group provides an excellent vehicle for learning and cross-training among group members and for development of individual members as well as the organization:

◎ **Individual** development within an organization must be reinforced in a team or group, or new skills will quickly be forgotten from lack of support on the job.

◎ For the same reason, **group/team** development within an organization must be pursued as a group working on skills together.

◎ **Organization** development—development of an entire firm, agency, division, or office—must be done in small, team-size groups to be effective.

If your organization has found either your training and/or your organization development efforts to be ineffective, you might want to read this (↑) again.

Individual Management Skills

The CEO of a small and growing design firm arranged for an annual retreat for the firm's management team. This group included all project managers, department managers, administrative managers, and financial and marketing managers (about 15 to 25 people in all).

Each retreat consisted of two basic tasks: business planning for the coming year and learning a needed skill together as a team. Over the years, they addressed skill development in staff productivity, conflict management, contract negotiations, management of meetings, and project planning and scheduling.

It was clearly beneficial to have carried out these retreats as a complete team. Because the group reinforced one another's individual skills, there was a significant improvement in those skills each year. Improvements were reflected in their clients' appreciation and in their profitability.

Firm-Wide Project Management Skills

The 20 most senior project managers of a national multioffice engineering firm gathered for two full days. Their tasks were to

◎ Pool their individual project management secrets and best practices around a central outline of 10 project management topics (e.g., scoping, pricing, negotiating, scheduling)

◎ Work together to develop a draft project management guidebook of 10 chapters, to be completed later by a single editor

◎ Propose a one- to two-day in-house project management seminar for less-experienced or new project managers, using the new guidebook as a resource

This firm recognized that the best experts on project management for their firm's unique needs were their **own** most senior and respected project managers. The guidebook and seminar they developed were outstanding for that firm and highly regarded by users.

Firm-Wide Strategic Planning

The operating committee of a construction firm prepared itself for a strategic planning session of several days. Rather than limit it to their own wisdom and participation, they involved other key leaders and clients:

◎ Each operating committee member convened several of his or her staff team with a few clients for a focus group.

◎ Each focus group identified clients' current perceptions and future needs and hopes.

◎ The operating committee members shared a summary of their focus group feedback prior to the planning retreat.

The resulting strategic plan had broader and better input. It was also more widely supported because other groups contributed.

CHAPTER 18

TEAM VERSUS INDIVIDUAL RECOGNITION AND REWARDS

> *The biggest hazard with **rewards** of all kinds is that leaders and managers **believe** they understand them. People are unique in most aspects, especially in what they see as rewarding!*

Nothing about leadership and management is as complicated and unfathomable as rewards, both tangible and intangible. Many firms make large layouts of cash, year after year, for very little in return.

Following are some conclusions I can safely draw from my practice and years of work with people in all areas of business:

★ Different people are motivated by different rewards, some of which don't even seem rewarding to other people.

★ Over the long haul, organizations attract the kind of people who are interested in the rewards they offer.

★ Being part of a good team is rewarding in and of itself, and all teams have norms about the rewards they share.

First, Some Theory

Many studies have shown that **people are motivated by very different things**. Motivation researcher Abraham Maslow noticed that people have a hierarchy of needs, beginning with survival needs and topping out with the reward of the work itself, called self-actualization.

Scientist Frederick Herzberg found that certain things are really "motivators," such as achievement, recognition, responsibility, and advancement. He also found that certain "satisfiers" were desired but actually did little to motivate; these include salary, supervision, policy, and coworkers.

Both Herzberg and Maslow have often been misunderstood as thinking that compensation is unimportant. Maslow would probably say that compensation is tied directly to some of the five basic needs people have (and is therefore important). Herzberg sees compensation as a satisfier (rather than a motivator) only because of poor administration practices, not because pay is unimportant!

Compensation research projects inside industrial and technical organizations have clearly shown that money and nondeferred bonuses and benefits actually do provide strong motivation for most people to perform. That is, **if** (and only **if**) the individual sees that his or her incentive pay is clearly and closely dependent on job performance, then he or she will be motivated.

Elements that are needed to improve the administration of salaries or bonuses include

★ clear-cut performance goals,

★ workable performance measures,

★ honest performance appraisals,

★ and then paying for performance and contribution only.

However, the way that bonus or incentive pay is usually administered does not follow this model. At least, many at the receiving end of compensation don't see it this way! Managers sometimes manipulate their pay-for-performance system to try to keep everyone happy. The result is that no one is happy and pay is not seen as related to performance.

More Recent Work

Perhaps the most extensive work on rewards in the form of compensation (salary, bonuses, and profit sharing) has been by Edward E. Lawler III. His book, *Pay and Organization Development* (Reading, MA: Addison-Wesley Longman Publishing Co., 1981; www.awl.com) presents critical findings with regard to "pay for performance" and how to do it well.

A controversial but valuable view is offered by Alfie Kohn in his book, ***No Contest: The Case Against Competition*** (Boston: Houghton Mifflin, 1992). Rewards, Kohn argues, encourage people to focus narrowly, do jobs fast, take few risks, and settle for lower-quality work.

Judith Bardwick, in her book, ***In Praise of Good Business: How Optimizing Risk Rewards Both Your Bottom Line and Your People*** (New York: John Wiley & Sons, 1998; www.wiley.com), makes the case for and against "security." She points out that **automatic** rewards are seen as entitlements, with **no** motivational utility or value!

Team Incentive Rewards

Several progressive firms, primarily in consumer goods, have experimented with and used team bonuses or incentives instead of individual incentives. They are concerned that individual incentives hinder or undermine teamwork.

That's a good news/bad news scenario. On the pro side, team incentives put team peer pressure on poor performers to shape up and on learners to develop quickly.

On the con side, team incentives don't necessarily recognize differences in skills or productivity among members and may frustrate the team's top performers and pacesetters.

Several consulting and design firms have developed what may be a practical compromise. The math is cumbersome but the idea has merit as seen by firms using this approach:

★ A portion (e.g., 50%) of the incentive pool is dependent on the individual participant's efforts and results relative to others.

★ The second portion is based only on relative salary levels (to reflect level of responsibility in the firm and to promote and support teamwork).

Does It Matter?

Given the varying views, new and old, one might well ask: "Does motivation make any difference?" My work with technical organizations of all kinds suggests that it does indeed make a large difference. Although it is difficult to measure productivity, particularly for knowledge workers, managers can assess relative output. As they have done it, particularly in groups, they have noted the following:

> Groups or individuals with high morale and high performance (with equal skills and resources) will outperform marginally acceptable groups or individuals by a factor of almost three to one.

Motivation makes a big difference in group output, from marginal workers to high performers (*Figure 18.1*). **Morale follows a parallel** rise from marginal to high.

Individual or Group Output as a Percentage of their Potential Output

100% --

   Burnout zone: May result in absence or illness

85% ---

   Optimal zone: High output, morale, teamwork

70% ---

   Average zone: Where many groups perform

50% ---

   Marginal zone: Poor output, low morale, little teamwork

30% ---

   Unacceptable zone: Managers don't tolerate this level of
                          output for long

0% --

Figure 18.1. Knowledge Workers and Groups: Relative Output Levels

This chart is based on empirical data developed from work with many management groups in consulting engineering, design, research, technology, construction, and public service.

Real Feedback Helps and Rewards

Real feedback means comments, praise, and suggestions from consumers, users, client groups, and field sales or service staffs. These comments may come through the leader or manager, but they need to come from the recipients of one's work, service, or product and they need to be specific, timely, and focused on performance goals and results. Several progressive firms in many business arenas have developed novel ways to get user and client feedback (see Chapter 12).

Most people generally want to make a contribution, to do useful needed work of value for other people—to serve others. That's where feedback comes in. **When they know they've made a contribution, people have a sense of achievement**; and that is an excellent all-around motivator.

Achievement leads to motivation, not the other way around. Running a close second is recognition for achievement. Recognition can take many forms, limited only by the leader's imagination.

Recognition and Reinforcement

Reinforcement makes use of the observation that many people want to be appreciated for what they're doing or at least simple acknowledgment from those they care about—leaders and team members. This is important because if work activity has no consequences when it occurs it will tend to die out over the long term.

Simple verbal recognition and acknowledgment in organizations is the greatest untapped, underused, effective, yet inexpensive method of rewarding and motivating people. It is underused by a factor of 10 or more in most organizations.

So What Do We (Almost) Know for Sure?

★ Different people are motivated by very different things.

★ Some motivators are tangible (like bonuses); many (like achievement) are not so tangible.

★ Organizations attract the people (long-term), who want the type of rewards that they offer.

★ Being part of a good team is highly motivating in and of itself.

★ Teams have norms (rules) about the rewards they share and value.

★ Pay can be a strong motivator when administered appropriately, as in "pay for performance."

★ Some component of pay might be for teamwork as well as for performance.

★ Motivation makes a big difference in the performance of a unit or individual—a potential of three to one (motivated versus marginal performance).

★ Client or user feedback is both rewarding and helpful in improving performance.

★ Simple acknowledgment and appreciation, when earned, are powerful for many people.

★ Other simple and sincere forms of recognition are also useful means of motivation.

CHAPTER 19

PROJECT TEAM LEADERSHIP

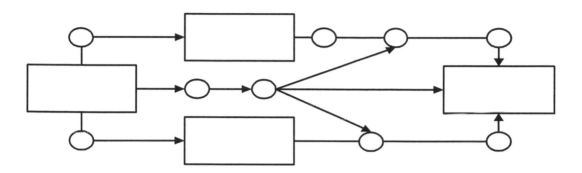

In the research lab, design office, fabricating shop, or on construction site, it is the **project manager** (PM) who must pull diverse talents and personalities into a project team.

Depending on the organization, this role is often called program manager, task force manager, or project team leader. Few would argue that the project manager is crucial to almost any successful endeavor.

The position description or responsibility list for such positions is typically twice as long as most other management roles, and includes

→ Scope   → Personnel
→ Cost   → Quality
→ Schedule   → Client satisfaction

Many research, manufacturing, design, and construction organizations with whom I've worked consider the project manager to be the most important management position they have.

Project Manager at the Center

Project manager is usually a pivotal position, **a key link among several groups**, including the project team, the client or user (whether internal or external), senior management, and in some cases the functional groups as shown in *Figure 19.1*.

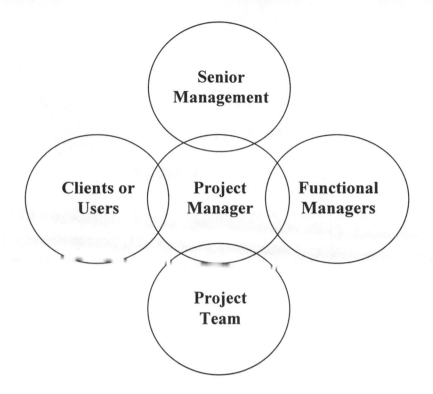

Figure 19.1. Project Manager as Key Link

As the link pin, the project manager's role is quite different in relation to each of these four groups. We can begin to see that the project manager must be a very skillful integrator of the efforts of others yet often without any clear authority over other contributors.

Here are some dimensions of the integrative roles of project managers:

→ The project manager must effectively represent the client's or user's interests to the project team, senior management, and functional management and must also satisfy the project requirements.

→ Functional management often exists alongside project management (as in a matrix structure). The project manager then shares the management of team members (who may also report to functional managers).

→ The project manager must try to balance what appear to be conflicting goals and priorities, including

  → work quality versus cost control

  → staying on schedule versus best use of the staff

  → satisfying the client and/or user versus profitability of the firm

The Project Manager and Project Team

The PM guide of a national transportation engineering firm states that project success, profitability, and client satisfaction are in direct proportion to the quality of project leadership in any organization. They define team leadership simply as the ability to get the job done through the effective and coordinated efforts and skills of others.

The PM responsibilities of a national environmental engineering firm contain over 20 items including the following **team-related responsibilities:**

→ **Leading** the project team by example

→ **Integrating** diverse disciplines and unique people

→ **Effective use** of the project team contributors

→ **Appropriate delegation** to team members, including follow-up

→ Establishing and maintaining clear and continuous **team communication**

→ **Proactive team building** for project teams

(The other important items deal with project planning, estimating, negotiating, marketing, quality control, technologies, representing the organization, accounting, billing, client communication, schedule attention, and development of people.)

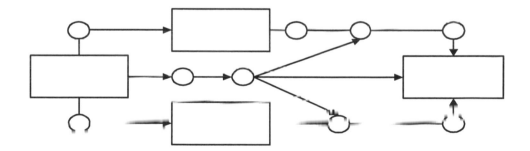

ASCE's workshop on **quality in the constructed project** identified a number of key factors to help ensure successful projects. Three are highly dependent on the project manager's leadership of the team:

→ The project manager must take the lead in developing guidelines and procedures for communicating the owner's objectives and expectations to the entire project team and for monitoring and ensuring satisfaction of those objectives and expectations.

→ The project manager must ensure that each engineering discipline represented on the design team has an understanding of the importance of the other disciplines and the value of adequate technical interfacing by all parties.

→ The project manager must strive for the synergism arising from multidisciplinary design review of the project—an important component of quality in any design/construction project.

To do this well, PM's must be quite effective at getting results through people, some of whom may not report directly to them!

Getting Results Through Others

Bohlen and colleagues[8] from the University of Dayton studied the strategies of PM's from the United States and Europe. They found that those PM's used the following strategies to get results:

1. **Consulting**—Gaining the support of team members by seeking their participation in the planning and decision-making process for a proposal or a project

2. **Reasoning**—Using factual information about a project to gain understanding, support, and assistance

3. **Relationship**—Building rapport and good relationships with team members before making a request

4. **Common vision**—Presenting project objectives in a way that appeals to team members' basic values and interests

5. **Enlisting support**—Developing or enlisting the support of other project contributors toward the project objectives in order to influence a particular member

6. **Higher management**—Using higher management authorization and support to directly influence team members

7. **Asserting**—Using assertive approaches such as clearly stated expectations for assistance to convince team members to do what's needed

---

[8] George A. Bohlen, David R. Lee, and Patrick J. Sweeney, **"A Comparison of U.S. and European Project Manager Decision-Making Styles,"** *Engineering Management Journal*, Vol. 7, No. 3, pp. 25–32, September 1995.

8. **Reciprocity**—Offering an exchange, such as reciprocal benefits, for doing what is needed on a project

9. **Forcing**—Threatening a team member with a negative outcome or penalty if the member does not assist with project needs and objectives

Different project contributors may and do respond to different strategies, and effective project managers must be able to use any of the nine strategies.

The Time Factor

In many services and organizations such as consulting, construction, engineering, research, development, design, and marketing, project teams often have special challenges:

→ They may be relatively short-term teams that must get up to speed quickly.

→ Members are sometimes added as work builds to a peak, then staff is pared down again.

→ In some situations, team members may not even be located at the same site.

These challenges can be addressed by the same types of team development tools and efforts discussed earlier; they simply must be more focused, more intentional, and faster.

Working with Functional Management

Many organizations that use project management also have functional groups or teams from which project team members are drawn. Sometimes team members stay with their particular functional groups; others come together in the project team location.

When team members come together with the project manager in the same location, the project manager's job is a bit simpler with respect to communication and teamwork. He or she may still rely on functional management for consulting expertise, quality assurance, and technical oversight as needed.

When project team members are **not** located with their functional or specialty groups, project managers will find the suggestions in Chapter 15 of value.

CHAPTER 20

PRECAUTIONS FOR PITFALLS AND PRATFALLS

> *It must be remembered that there is nothing more difficult to plan, more uncertain of success, nor more dangerous to manage than the creation of a new order of things. For the initiator has the enmity of all who would profit by the preservation of the old institutions, and merely lukewarm defenders in those who would gain by the new ones.*
>
> —Niccolo Machiavelli's advice to the leaders of his day[9]

While the team approach is neither fragile nor risky, there are some times and places where pitfalls may exist and precautions should be taken. In many cases, these pitfalls and precautions exist whether or not a team approach is used. The point is that a team approach doesn't necessarily remove the pitfall.

An exhaustive list would probably cover 200 situations; here we plan only to consider those that are quite common:

⬦ The **continuous tension** between helping and hindering forces

⬦ **Changing from a traditional group** to a team approach

⬦ The team approach in a **nonteam corporate culture**

⬦ **Considering a "matrix"** type of organization

⬦ **Setting policy** by setting precedents

⬦ The **need for discipline** and disciplinary steps

Helping and Hindering

At any point in time, a team will have certain forces helping teamwork and other forces making teamwork more difficult. These forces reach a sort of equilibrium for whatever level of teamwork exists at the time. Next week, the forces will probably change, and so will the balance point *(Figure 20.1)*.

---

[9] *The Prince*, 1513.

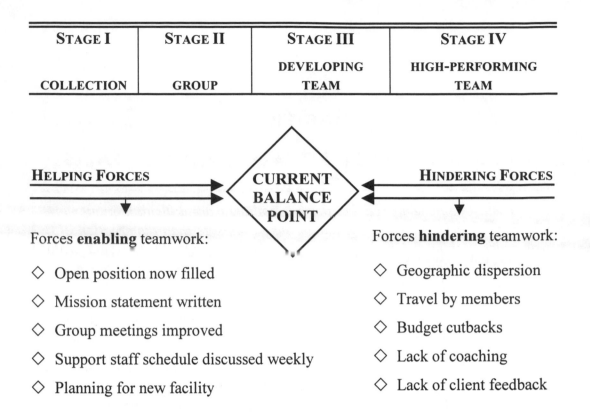

STAGE I	STAGE II	STAGE III	STAGE IV
COLLECTION	GROUP	DEVELOPING TEAM	HIGH-PERFORMING TEAM

HELPING FORCES → **CURRENT BALANCE POINT** ← **HINDERING FORCES**

Forces **enabling** teamwork:

◇ Open position now filled

◇ Mission statement written

◇ Group meetings improved

◇ Support staff schedule discussed weekly

◇ Planning for new facility

Forces **hindering** teamwork:

◇ Geographic dispersion

◇ Travel by members

◇ Budget cutbacks

◇ Lack of coaching

◇ Lack of client feedback

Figure 20.1. Balance of Forces

The world generally works to provide plenty of difficulties and forces that will and do make teamwork more difficult. "Entropy" sets in and the team runs down. Team leaders and members need to find ways to create more helpful forces that will enable and support teamwork. It's a never-ending task. You can never rest on your laurels!

Traditional Group-to-Team Approach

Many newer and younger managers are often interested in taking more of a team approach to managing their unit, group, or organization. Troubles can bubble up, however, when such a manager, through promotion or hiring, is given responsibility for a traditionally managed group.

A traditionally managed group is one in which the previous manager took a directive approach to managing. That is, he or she organized, assigned, controlled, monitored, and corrected the people and the systems almost exclusively by him- or herself.

The problem in this case is that the staff of such a group has been trained to leave such matters up to their manager. The new manager may expect the group to welcome involvement in planning and problem solving. Instead, the staff may be quiet, passive, and unenthusiastic. They might even complain about this new expectation of their new manager.

The transition from a typical group approach to a team approach can take some time. The adjustment requires both new attitudes and skills, something the new leader must take the lead in developing. This sort of culture change can take up to two years, even with strong effort by an excellent manager, but it's usually worth the time and effort!

Team in a Non-Team Culture

Many smaller firms and agencies have moved toward more of a team culture because of a courageous and persistent manager somewhere down in the ranks. He or she develops a high-performing team in a nonteam culture, and it spreads like a positive virus. Other managers, even senior management, may adopt it because they like what comes with it.

The possible pitfall with a **team** approach in a **non**team culture is that other managers may not like it. They may be perturbed because the group operating as a team is less dependent on its manager. Or they may fear it as an uncomfortable expectation for themselves.

In situations in which other managers feel threatened, it's relatively easy for them to paint a "negative picture" of the team approach:

◇ Things seem to be "out of control."
◇ The boss doesn't "have all the answers."
◇ Communication is frequent and straightforward.
◇ Issues are openly discussed, and people disagree strongly at times.

The team approach is easiest for others to badmouth when it's in stage II. The best strategy for avoiding negative attention is to move the team rapidly into stage III or IV, where it's much harder for others to find fault with the team approach.

Setting Policy by Setting Precedents

During the first few months of any new system, program, policy, plan, or operation, the staff is discovering the real rules of the game. They also pay attention to who means what they say and what the policy really is versus what's posted or written in the employee handbook.

Most people recognize that there are two kinds of policies }	Written and/or official rules and policies	*and*	Policies set by practices and precedents

During the initial period of any new system or program, managers must be careful to get people started right. A bit of care and courage during this time can save a lot of hassles later. Taking things away later or letting bad practices continue is much harder than doing it right from the beginning.

A similar thing happens when a new employee joins an existing work group. He or she will try to figure out the real "rules of the game" from fellow employees. During this period, the manager's orientation, coaching, and cues are most important.

Discipline and Disciplinary Steps

As they begin to understand, appreciate, and move toward more of a team approach to the leadership of their staffs, leaders sometimes stumble in the area of discipline. Perhaps they assume that once everyone is tuned into a team approach, problems like tardiness, poor attendance, poor work habits, grumpiness, or worse, will disappear. Although they often do, that's not always the case. Some employees may require expert counseling as well as the attention of management.

Most employees want to make a contribution. The only discipline they will ever need during their careers is training and coaching. However, managers may need to use appropriate disciplinary action from time to time, and it must not be avoided when needed. **For those relatively few times when discipline is important, it is very important.**

Discipline can be defined as

◇ A minimum, consistent, and reasonable system of rules and procedures for the conduct of members and performance of their work

◇ Training and coaching that develop self-control, character, orderliness, safety, and efficiency on the job

◇ Methods of problem solving when situations, behaviors, or performance are not what they need to be

The need for disciplinary action is less likely if

◇ The leader sets a good example in his or her own work.

◇ Staff training for tasks and procedures is adequate.

◇ The leader sets clear and high expectations or enables the staff to set them as a team.

◇ The employee's development is observed and helped if needed.

◇ Positive consequences are provided for high performance, including recognition, respect, and responsibility.

◇ Minor problems are discussed **when** they occur, and the employee is encouraged to solve them.

Criteria for **fair disciplinary actions**, taken from arbitration cases, include the following:

◇ The employee has had **forewarning** or foreknowledge of the possible disciplinary consequences of his or her conduct.

◇ The rule or directive was reasonably related to (a) orderly, efficient, and **safe operation** and (b) **performance** properly expected of employees.

◇ **Before taking disciplinary action**, an effort was made to discover whether there was, in fact, a problem.

◇ Rules, directives, and **penalties have been applied evenhandedly** and without discrimination, to all employees.

◇ Any disciplinary action in a particular case was reasonably related to (a) the **seriousness of the event** or problem and (b) the **past record** of the employee.

Making Disciplinary Actions Productive When They Are Needed

These days many firms and organizations have their own policies and procedures for promptly handling employee problems and maintaining morale, service, productivity, quality, and teamwork. Here are some basic steps often recommended by human resource professionals when regular performance counseling alone hasn't improved an employee's problem of performance or teamwork.

1. **Make expectations clear.** It's helpful to describe the employee's actual work performance with reference to needed and expected performance (or teamwork). Encourage the employee to respond and explain. Probe for full information and circumstances on the matter.

105

Be specific in the results you need, and mention the importance of their work to the firm. Check to be sure the employee understands and makes a commitment. Most importantly, express confidence that the employee can improve. A positive expectation is essential here, as it is at most times.

Bear this thought in mind: Many potentially fine employees are shocked and startled to (finally) learn about significant performance or teamwork problems for the first time when they are discharged. Usually, managers think they have been very clear about expectations when, in fact, they have been vague, brief, or even confusing!

2. **Review Expectations.** This is another meeting to review your needs and desires for the job regarding performance and teamwork.

   If improvement has been inadequate, it may be helpful to honestly discuss that fact and to give some detailed information as to what is still missing.

   If improvement has been significant, it's vital to verbalize that, again being specific. Whatever you recognize and acknowledge will probably be further improved.

3. **Give the employee time to think it over.** A growing number of organizations have found that suspension with pay has the positive factors of reducing anger and hostility about lost pay while also giving the employee a highly visible notice of the seriousness of his or her situation. Again, it is very helpful to review the problem and previous discussions about expectations and the needs of the job and the team.

   Ask for and listen carefully for any new circumstances or facts from the employee. Advise the employee to take tomorrow (or however long is appropriate) off from work. He or she is to go home in order to think about his or her decision "because the time has come for you to decide whether you wish to continue working with us." Express your hope and positive expectation that he or she can change.

   This step is a very positive and humane way to let the employee know the seriousness of his or her situation, encouraging them in the strongest possible way to realize the potential for him- or herself and the team.

4. **Positive outplacement:** These days, more firms of all kinds are recognizing their obligation to assist employees in refocusing their work and lives when discharge is necessary for any reason, including reductions in staff, poor performance, or other problems.

The most readily available avenue for assisting the discharged employee is through outplacement counseling by an individual or a firm specializing in this service. The service can be as brief as assistance with emotional adjustment and job-finding efforts such as networking, resume writing, and interview skills; or it may include help with job hunting, having a temporary work office address, and other counseling as needed.

Pleasant surprise—this kind of care and help often has its greatest possible impact on the remaining team members! It helps them move past any feelings of guilt or anger for the firm's actions. Of course, the real service is to the discharged employee, who gets practical help in restarting his or her work life. Those employees, when well counseled, often report greater job satisfaction and life fulfillment as a result of these necessary changes!

CHAPTER 21

RESOURCES

For those who want to gain still more insight, skill, or versatility on teams, team leadership, and management, the following publications may be helpful in addition to those mentioned throughout the book.

Books

📖 *The Conflict-Positive Organization: Stimulate Diversity and Create Unity*, a book by Dean Tjosvold about the naturalness of conflict and the danger of avoiding it, published by Addison-Wesley Longman, Reading, MA, 1991; www.awl.com.

📖 *Continuous Excellence*, by Mel Hensey, a handbook for organizational improvement—and teams are part of that; published by ASCE Press, Reston, VA, 1995; www.asce.org, 800/548-ASCE.

📖 *Designing and Leading Team-Based Organizations*, a workbook for organizational self-design by Susan and Allan Mohrman, Jr., published by Jossey-Bass, San Francisco, 1997; www.josseybass.com.

📖 *For Team Members Only*, a practical handbook by Manz, Neck, Mancuso, and Manz for team members, published by AMACOM Books, New York, 1997; www.amanet.org/books.

📖 *Heart of the Mind: Engaging Your Inner Power to Change with Neuro-Linguistic Programming*, a book by Connierae and Steve Andreas about personal change and resources, published by Real People Press, Moab, UT, 1989; www.realpeoplepress.com.

📖 ***High Involvement Strategic Planning***, a book by Robert G. Cope about leadership and participative planning, published by The Planning Forum, Oxford, OH, 1991.

📖 ***Lincoln on Leadership***, a book by Donald T. Phillips on Lincoln's overlooked qualities as a great leader, published by Warner Books, New York, 1993.

📖 ***Making Teams Work***, a handbook for teams and team building, published by Organizational Dynamics, Inc., Billerica, MA, 1993; www.orgdynamics.com, 800/ODI-INFO.

📖 ***Managing in the New Team Environment: Skills, Tools and Methods***, a book by Larry Hirschhorn about team structure and process, published by Addison-Wesley Longman, Reading, MA, 1991; www.awl.com.

📖 ***New Ways of Managing Conflict***, a book by Rensis Likert and Jane Gibson Likert about organizations operating on four different systems, particularly system 4, published by McGraw-Hill Book Co., New York, 1976; www.mcgraw-hill.com.

📖 ***Partnering for Success***, a booklet by Thomas R. Warne, published by ASCE Press, Reston, VA, 1994; www.asce.org, 800/548-ASCE.

📖 ***Personal Success Strategies: Developing Your Potential*** by Mel Hensey, has several sections appropriate to team members, conflict resolution, and more, published by ASCE Press, Reston, VA, 1999; www.asce.org, 800/548-ASCE.

📖 ***Power Up***, a book about organizational transformation through shared leadership by David L. Bradford and Allan R. Cohen, published by John Wiley & Sons, Inc., New York, 1998; www.wiley.com.

📖 ***Skill Building for Self-Directed Team Members: A Complete Course***, a handbook for team members by Ann and Bob Harper, published by MW Corporation, Mohegan Lake, NY, 1992.

📖 ***Succeeding as a Self-Managed Team***, a helpful guidebook to self-led teams by Chang and Curtin, published by Richard Chang Associates, Inc., Irvine, CA, 1994; www.richardchangassociates.com, 800/756-8096.

📖 ***The Team Handbook: How to Use Teams to Improve Quality***, 2nd ed., a spiral bound team handbook focusing on quality improvement by Peter R. Scholtes, Brian L. Joiner, and Barbara J. Streibel, published by Oriel, Inc., Madison, WI, 1996; www. orielinc.com, 800/669-8326.

📖 *Vroom! Turbo-Charged Team Building*, a management comic book by Michael Shandler and Michael Egan, published by AMACOM Books, New York, 1996; www.amanet.org/books.

Journals and Newsletters

📖 *Journal of Management in Engineering*, a quarterly refereed journal published by ASCE Press, Reston, VA; www.asce.org, 800/548-ASCE.

📖 *Leadership and Management in Engineering*, a quarterly practice-based journal published by ASCE Press, Reston, VA; www.asce.org, 800/548-ASCE.

📖 *Leadership Strategies*, a monthly newsletter with lots of practical tips on leadership and teams, published by Leadership Strategies, Washington, DC; 800/915-0022.

📖 *Today's Team*, a monthly newsletter published by Wentworth Publishing Co., Lancaster, PA; 800/822-1858.

INDEX

Other Books by the Author ...

CONTINUOUS EXCELLENCE:
BUILDING EFFECTIVE
ORGANIZATIONS
(ASCE Press, 1995)

and

PERSONAL SUCCESS
STRATEGIES: DEVELOPING
YOUR POTENTIAL
(ASCE Press, 1999)

Available from ASCE
Call 800/548-2723 (US)
703/295-6300 (Int'l)
http://www.pubs.asce.org